Optoelectronics

TUTORIAL GUIDES IN ELECTRONIC ENGINEERING

Series editors
Professor G.G. Bloodworth, *University of York*
Professor A.P. Dorey, *University of Lancaster*
Professor J.K. Fidler, *Open University*

This series is aimed at first- and second-year undergraduate courses. Each text is complete in itself, although linked with others in the series. Where possible, the trend towards a 'systems' approach is acknowledged, but classical fundamental areas of study have not been excluded. Worked examples feature prominently and indicate, where appropriate, a number of approaches to the same problem.

A format providing marginal notes has been adopted to allow the authors to include ideas and material to support the main text. These notes include references to standard mainstream texts and commentary on the applicability of solution methods, aimed particularly at covering points normally found difficult. Graded problems are provided at the end of each chapter, with answers at the end of the book.

Optoelectronics

J. Watson
Senior Lecturer
Department of Engineering
University of Aberdeen

 Van Nostrand Reinhold (UK) Co. Ltd

First published in 1988 by
Van Nostrand Reinhold (UK) Co. Ltd
Molly Millars Lane, Wokingham, Berkshire, England

Typeset in Times 10 on 12pt by
Colset Private Ltd, Singapore

Printed and bound in Hong Kong

British Library Cataloguing in Publication Data

Watson, J. (John)
    Optoelectronics. — (Tutorial guides in electronic engineering; 14).
    1. Electrooptics
    I. Title   II. Series
    535   QC673
    ISBN 0-278-00008-8

ISSN 0266-2620

Contents

Preface

The intention of this introductory text is to stimulate the interest of engineering students in optoelectronics and provide, through a systems approach, an insight into some of the existing developments being made in the field. In a book of this nature, it is impossible to provide a comprehensive treatment of so diverse a subject and, of necessity, the chosen subject material largely reflects my own personal opinion of what constitutes the essential material of an optoelectronics course. It is hoped, however, that the coverage here will serve as a useful introduction to the subject and provide the basic framework upon which the reader will be able to build in later years, either through taught instruction or by his own efforts.

The format of the book reflects the systems nature of the treatment. Chapter 1 is devoted to describing the basic elements required of an optoelectronic system and illustrating their importance through the medium of a case study. The particular case study chosen, the optical compact disc system, epitomizes the optoelectronic design philosophy and highlights the significant advantages which can often be afforded by choosing light as the medium by which we carry information. From this introduction, the individual elements are discussed in turn, starting with, in Chapter 2, an outline of the basic concepts of light emission from gases and solids and how these processes can lead to laser action in a given medium. Chapter 3 takes these concepts and discusses the practical constraints necessary to produce working lasers together with a discussion of the operation and characteristics of some of the more common lasers encountered in industry. No attempt has been made to make this an exhaustive coverage of every laser available. The aim has been more to describe one or two of the basic types while leaving room for the reader to find out more for himself.

The treatment of photodectors is, similarly, not intended to be exhaustive but more an introduction to the principal methods of light detection employed in optoelectronics. Chapter 5 brings together the two principal elements in the optoelectronic chain, namely the source and detector, and outlines the principles of optimum light coupling between the two. This section is crucial to good optoelectronic systems design but has generally been omitted from introductory texts on the subject.

Finally, in Chapter 6, we discuss some typical optoelectronic systems and applications. The applications have been chosen to give an insight into some of the novel and diverse applications of optoelectronics in industry. To this end, all the examples are actual industrial case studies and, in all cases, the techniques are currently employed in industry. The method of approach has been not to give a global overview of an entire field but to select a particular system which illustrates the concepts and design philosophy adopted by the engineering team. For example, rather than attempt a definitive coverage of optical communications I have chosen to outline some actual optical links in use in the British Telecom trunk services network. It is hoped that this case study approach gives the reader a better awareness of engineering design concepts.

No text nowadays can be written without a lot of help, forbearance and understanding from others: this book is no exception. Amongst the people I am indebted to are friends and colleagues of long-standing: Drs Bryan Tozer and Steve Adrain of Marchwood Engineering Laboratories for supplying me with detailed information and photographs of some of the optoelectronic work currently underway within the CEGB and for constructively criticising sections of the manuscript; Mr John Wilson of Philips Electronics for providing me with so much useful information on the CD system; Professor Alan Rogers of King's College, London, for taking the time to discuss optical sensors with me; Mr Martin-Royle of British Telecom for supplying much of the information relating to the use of optical fibres within the UK telephone network; Professor Kel Fidler for keeping a watchful and encouraging eye over progress; the CEGB for permission to use modified diagrams; Bryan Tozer of CEGB Marchwood Engineering Labs for photographs and diagrams. Thanks are due to Osprey Electronics, RCA, Radiospares, Lumonics (JK Division), Quantel, Spectra-Physics and Coherent (UK) for permission to publish photographs of and data on some of their products. I wish also to thank my colleagues at Aberdeen University who have made helpful suggestions on the manuscript and my postgraduate students, past and present. Lastly, and definitely not least, my thanks to Joyce, Jane, Ewan, Morag and Fiona.

Elements of Optoelectronics   1

Objectives

☐ To define what we mean by optoelectronics
☐ To discuss its impact on our lives
☐ To outline the principal elements required in an optoelectronic system
☐ To illustrate the design concepts of optoelectronics by an analysis of the compact optical disk system.

Having only recently risen to prominence as a scientific discipline in its own right, *optoelectronics* is rapidly becoming established as one of the corner-stones of modern engineering. Drawing as it does on aspects of electronics, optics, electromagnetism and the science of materials, optoelectronics can be regarded as a true multi-disciplinary subject which is of relevance to all students of engineering be they electrical, electronic, mechanical or civil. It is this diversity of interests which is the strength of optoelectronics and makes it such an exciting and absorbing field of study. The purpose of this book is to instil in the reader an awareness of the scope of optoelectronics, provide an insight into the many exciting developments being made and outline the basic skills required to design simple optoelectronic systems.

Other books, that you should find of interest include: Wilson, J. and Hawkes, J. *Optoelectronics: An Introduction* (Prentice-Hall 1983); Seymour, J. *Electronic Devices and Components* (Pitman 1981); Chappell, A. *Optoelectronics: Theory and Practice* (Texas Instruments, 1978).

What is Optoelectronics?

Although, as we have said, optoelectronics is a relatively recent addition to the range of skills required by the practising engineer, its roots can be traced back to the time when man first attempted to use natural light as a means of measurement or communication. The signal fire and the sundial were perhaps the first 'optoelectronic' devices.

This may seem like stretching credibility a bit, but if we think about the nature of light we can see that optics and electronics are linked through the structure of the atom itself.

We will return to this theme in Chapter 2.

From these early beginnings, optoelectronics has grown through the invention of the incandescent lamp, the cathode ray tube, the transistor, the laser, the optical fibre and the microprocessor to become one of the most significant developments of the technological age. Its impact on our working and everyday lives is easily seen.

Here I have singled out a few inventions which I personally regard as some of the milestones in the development of optoelectronics.

In engineering, low-power lasers are used in the alignment of buildings, holography is used to measure surface displacements of vibrating machinery to sub-micron precision, sensitive light sensors allow robots to 'see', while optical computers offer unparalled processing power. In our everyday lives, the impact of optoelectronics is equally dramatic. Telephone conversations are transmitted by light through lengths of glass fibre several kilometres long, television allows us to monitor events around the world, the compact disk player replays music by light and at the supermarket checkout holographic scanners record our purchases.

For further insight into the world of optoelectronics, read the popular scientific literature such as *Scientific American*, *New Scientist*, *Laser Focus*, *Lasers and Applications*, *Electronics and Power* and *Physics Bulletin*.

Exercise 1.1
Can you think of any more applications of optoelectronics in your lives, other than those mentioned above? For a start, look at your wrist-watch or think about the way you remotely control your television or even the way in which your books are checked out at the library.

The above examples, and more, all illustrate the immense scope of optoeletronic activity which ranges from photography to holography, from television to optical communications, from basic optics to laser physics, and demonstrate its transcendence of inter-disciplinary barriers. Optoelectronics in its widest sense represents the application of electronics to optical systems. By providing the designer with a radically new approach to system design, in which light is used as the means of carrying information and electronics to control and process it, opto-electronics may be set to revolutionize engineering in a way which could emulate that of the microprocessor.

In the pages which follow we will try to give as broad a view of the optoelectronic spectrum as possible and transmit a flavour of the excitement of this field. First of all though, we need to define what is meant by an optoelectronic system.

The Optoelectronic System

All the systems mentioned earlier comprise, in one form or another, a source of light, a detector and some means of relaying light between the two; add to this any power supplies and processing of the input and output signals which may be required and we have the basic elements of an optoelectronic system (Fig. 1.1).

Looking now at the individual elements in turn we can see something of the decisions we have to make in designing a particular system.

Starting with the light source our choice is wide. On the one hand we have the traditional *thermal* sources such as the gas discharge or tungsten-filament lamp or, on the other, one of the *new wave* sources like the laser or the light-emitting diode. It is true to say that the invention of the laser is the single most important event in the development of optoelectronics; for a long time a 'solution without a problem', the laser is now seen in every walk of life from laser-induced thermonuclear fusion to laser-light shows. In choosing a light source, some of the parameters which have

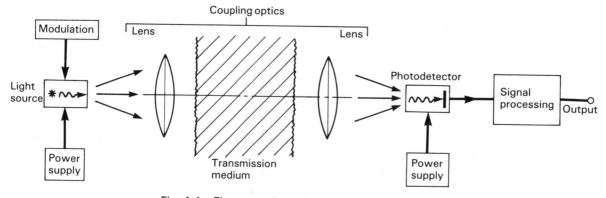

Fig. 1.1   Elements of a typical optoelectronic system.

to be taken into account include its radiant power, its spectral distribution and the way in which the light spreads out in space.

These parameters are discussed in more detail in Chapter 5.

Choice of photodetector is equally important as that of the light source. The human eye, although very sensitive, does not allow quantitative measurements to be made, nor can it permanently store information. To be able to display, measure or record the detected light signal, we have to consider the use of electronic detectors such as junction photodiodes, the TV camera and charge-coupled devices (CCDs). One extremely important detector which is all too frequently omitted from a discussion of photodetectors is photographic film. Although not an opto-electronic detector in the 'electronics' sense of the word, it offers high sensitivity, wide spectral range and permanent storage of information. It should not be disregarded. The final link in the optical chain is the relaying optics. In general we need to couple as much light as possible from the source to the detector while still retaining a high information content. These needs are influenced by the choice of coupling lenses, the transmission medium and the power and spatial distribution of the source light. Just which of these components we need in a system and how they are configured depends, largely, on what we require from it. In its simplest form a system might merely consist of sunlight and your eyes, with perhaps a neutral-density (e.g. sunglasses) filter between the two to stop you blinding yourself!

The human eye can detect a few-hundred photons! See Chapter 2.

Such a basic system, as that described above, enables us to go about our daily lives with amazing precision and detail. We can distinguish between day and night and a continuum of shades in between; we can see all the colours of the rainbow and detect subtle differences of hue; we can resolve dimensional detail down to fractions of a millimetre and can detect minute amounts of light. What we cannot do is quantify these phenomena readily or detect anything outside the visible spectrum. This is where optoelectronics comes in. By replacing sunlight with a man-made light source we can control the amount and spectral character of the light and by using a photodetector instead of the eye we can store and measure the quantity of light transmitted.

A neutral-density (ND) filter is a piece of glass or gelatin which absorbs light equally over a narrow wavelength band.

A simple system incorporating optoelectronic components consists of a light-emitting diode and a silicon photodetector on opposite edges of a garage door. When the beam path is broken the door opens. Can you describe, in simple terms, how this might work?

Exercise 1.2

We can illustrate some of the points we have been making as regards system design, by analysing the principles involved in the design of one of the most successful optoelectronic systems yet devised; the compact disk system.

The Compact Optical Disk (CD) System

The Compact Optical Disk system is, arguably, the epitome of the 'optoelectronics age': by offering a radically new approach to the recording and reproduction of audio information, the designers were able to approach the entire concept of music reproduction from a fresh standpoint.

In deciding to use light as the means of extracting information from the disk they were freed from the constraints imposed by trying to improve and modify the exist-

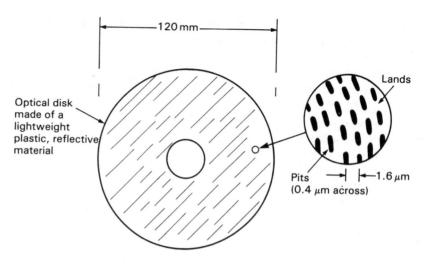

Fig. 1.2   The compact audio disk.

For a discussion of the principles of microprocessors see Downtown, A.C. *Computers and Microprocessors*, Tutorial Guide No. 4 (Van Nostrand Reinhold, 1984).

ing analogue recording technology and were able, from the outset, to contemplate a wholly digital recording and replay system. In fact, the creation of the optical disk system would not have been possible without the parallel developments of the laser and microprocessor.

The underlying principles of the CD system are simply stated. The key to understanding the system is the disk itself. Unlike a conventional black vinyl record which stores the audio information as an *amplitude modulated* groove, in which the lateral displacement of the groove is proportional to the intensity of the signal, the CD stores the signal in *digital* form by way of a spiral track of 'pits' of varying lengths embedded in a lightweight, reflective material and sandwiched between layers of plastic (Fig. 1.2). It is the pits and 'lands', the raised portions between the pits, which provide the digital coding of the signal as a series of 0's and 1's. A laser beam scans the track and processing electronics translate the variation in reflected light into a binary electrical signal which is subsequently decoded and converted back into its original analogue form.

Optical Readout

At the heart of the CD player is the optical pick-up. The purpose of the pick-up is to extract the digital information from the encoded disk and feed this to the signal processing section. Within the pick-up head are all the basic system elements we discussed earlier: the source, the photodetector and the relaying optics. The layout of a typical pick-up is shown in Fig. 1.3.

The pick-up shown here is the type used in the latest generation of players by Philips Ltd of the Netherlands.

One advantage in using a laser is that laser light is easy to collimate. See Chapter 2.

A semiconductor laser directs light towards the surface of the disk via a partially reflecting mirror and a collimating lens. Most of the light which strikes the partially reflecting mirror is reflected towards the disk although a small portion will be lost through absorption at the mirror surfaces and by transmission through it. The collimating lens directs parallel light into the focusing lens and so ensures that focusing is independent of beam path length. A second lens focuses the laser light onto the surface of the disk. Light which is reflected from the disk returns along the optical path and, this time, passes directly through the semi-reflecting mirror to a wedge-shaped lens and then on to the detector. The wedge lens splits the incident

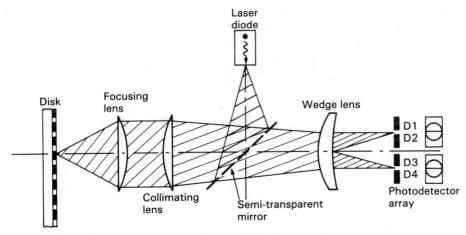

Fig. 1.3 Optical pick-up.

beam into two output beams which are directed towards separate double-photo-diode arrays to provide focusing and tracking information. The lens also corrects for any '*astimagtism*' introduced at the mirror. Astigmatism occurs when a wavefront displays uneven curvature. It manifests itself as a difference in focal position for rays in horizontal to those in a vertical plane. The lens corrects this defect by having a different focal length in the horizontal and vertical planes.

Three distinct mechanisms are responsible for the return of light falling on the surface of the disk: reflection, interference and diffraction. If the focused spot of light falls on a 'land' the high reflectivity of the disk results in nearly all of the light being reflected back along the beam path (Fig. 1.4). If, however, the spot falls partially or totally on a pit there will be a reduction in the intensity of the return beam due to the combined effects of interference and diffraction.

Interference is an optical phenomenon which occurs between two or more optical waves which meet in space. If these waves are of the same wavelength and

This unique lens was designed specially by Philips for their CD players.

For a full discussion of astigmatism, reflection, interference and diffraction refer to Hecht, E. and Zajac, A. *Optics* (Addison-Wesley 1974).

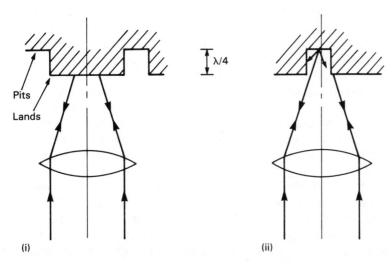

Fig. 1.4 Readout of information from disk. At reflective surface (i) most of incident light is returned. In pits (ii) light irradiance is lost by interference and diffraction.

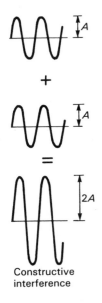

Constructive
interference

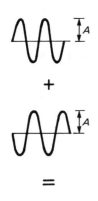

Destructive
interference

An important point is being
made here: when contemplating
the design of a new system, it
is the entire system which must
be evaluated not just individual
aspects.

meet such that the crest of one wave coincides with that of the other, the waves are said to be *in phase*. The peak amplitude of the combined wave is the sum of the amplitudes of the individual waves. This is *constructive interference*. Two waves will interfere constructively if they meet after travelling paths which differ in length by a multiple of one wavelength. The opposite effect, *destructive interference*, arises when the peak of one wave coincides with the trough of the other. The waves are one-half wavelength *out of phase*. The net effect is to cancel out any disturbance at that point. This effect can occur when the two waves meet after travelling paths differing in length by an odd multiple of half wavelengths. The depth of a pit on the disk is 0.11 μm, which after allowing for the refractive index of the transparent layer is about one-quarter of a wavelength of the illuminating beam. If the beam falls partially on both pit and land, a ray reflected from the bottom of the pit will destructively interfere with one reflected from the land, resulting in a decrease in intensity of the return signal.

The other phenomenon which influences readout is diffraction. This occurs when light spreads after passing through an aperture. It is only observed if the dimensions of the aperture are of the same order of magnitude as the wavelength of light which falls on it. Any part of the beam falling into a pit will be diffracted within it since the width of the pit is about the same as the wavelength of the laser beam and, instead of returning along its original path, is spread out over a much wider angle, again resulting in a decreased return signal.

Hence the radiant power of the reflected optical signal is modulated in accordance with the pattern of pits and lands on the disk. The resultant output of the photodetector is fed to the signal processor.

Let us now look at the specific components in more detail.

The Laser

At this point we should ask ourselves a fundamental question: *why use light to read the information?* It should be fairly clear that, because there is no mechanical contact between the pick-up and disk, using light to extract information eliminates wear and removes surface noise. However, using light for the readout of information also implies that light can be used in the recording of information. Since, as we will see later, light can be focused to small spots, vast amounts of data can be stored in a relatively small area of disk. It is probably these advantages of low noise and mass storage which justified the costly and time consuming development of the CD system. Additionally, using a spot of light under microprocessor control to track the disk eliminates signal distortion due to mis-tracking. Poor tracking of the stylus on the disk is one of the most common causes of signal distortion in conventional analogue disk systems. Furthermore, the absence of mechanical contact ensures that signal processing can be made independent of motor speed variations. The audible effects of speed variations, such as *wow* and *flutter*, can be almost eliminated.

Having decided that light is a sensible tool to use to extract information from the disk, then why choose a laser? Why not an LED or even a tungsten bulb? There are several reasons why a laser is the best for this application. Firstly, because we want to monitor the tiny proportion of light which is reflected back from the surface of the disk we need a source with a high-output flux. Secondly, to focus light to a small spot size requires that the source has a high degree of spectral purity and as short a wavelength as possible. The purity of a light source is related to its output spectrum: the narrower the bandwidth of the spectrum, the smaller is the focal spot size. A white-light source, by definition, encompasses all wavelengths from about 400 to 700 nm (the visible spectrum), whereas a light-emitting diode possesses a spectral bandwidth of about 50 nm centred on its peak-emission wavelength. A laser is the closest approximation we have to a monochromatic light source available, with bandwidths of 10 nm for a semiconductor laser and less than 0.1 m for gas lasers. Thirdly, to enable us to readily collimate the light and facilitate alignment we need a source with a low spatial spread. The spatial spread of light is, like its spectral bandwidth, related to the nature of the source. Incandescent lamps emit light into an approximately spherical distribution, whereas, a laser emits in an extremely narrow cone of light which is easily focused to a small spot. All these factors dictate the use of a laser and in particular a semiconductor laser because of its small size, efficiency and low power requirements. The actual laser used in the Philips system is based on a small crystal of aluminium gallium arsenide (AlGaAs), delivering about 5 mW of continuous power at a wavelength of 800 nm.

In principal, a monochromatic source is composed of a single spectral line with an infinitesimally narrow width.

Spectral and spatial distributions of light sources are discussed in more detail in Chapter 5.

A useful article on the requirements of a laser for optical disk readout is Van Ruvyen, L.J., A semiconductor Laser for Optical Disk Systems, *IEEE Transactions on Consumer Electronics*, May 1982.

We can see how the use of a laser helps us by estimating the number of bits of data we could record on an optical disk. The spot size we can focus a light beam to is dependent on the wavelength of the light and the numerical aperture (NA) of the lens used to focus it. For a wavelength of 800 nm and an NA of 0.5 the spot diameter d is,

Worked Example 1.1

The NA is related to the largest angle of light which can be accommodated by the lens. See Chapter 5.

$$d = \text{wavelength/NA}$$
$$= 800 \times 10^{-9}\,\text{m}/0.5$$
$$= 1.6 \times 10^{-6}\,\text{m}$$
$$= 1.6\,\mu\text{m}. \tag{1.1}$$

The area encompassed by such a spot is just $\pi d^2/4$. To a first approximation, the number of spots which can be packed onto a disk of diameter D is just the area of the disk divided by the spot area. Assuming that each spot corresponds to a bit of information, the number of bits, is given as,

The Post Office have compressed all 23.5 million private and business addresses in the UK onto a *single* CD read-only-memory disk produced by Hitachi.

$$\sim D^2/d^2$$
$$\sim (120 \times 10^{-3}\,\text{m})^2/(1.6 \times 10^{-6}\,\text{m})^2$$
$$\sim 6 \times 10^9 \text{ bits}$$

This is enough capacity to store this entire book several times over!

The above analysis shows us that the shorter the wavelength of light used the greater is the information density. Why then not use a HeNe laser, with its wavelength of 632 nm, rather than a semiconductor laser? In fact the laser video disk players introduced a few years earlier than the CD did use such a laser because of the greater information capacity required to store video signals. Since audio signals, however, require less in the way of storage bits and player compactness and low cost were included in the design brief, the diode laser was chosen for the CD system. A point to note though is that the video disk uses an analogue technique, pulse-width modulation (PWM), to encode the disk.

The video-disk format has been used as the basis of an all-electronic version of *Domesday Book*: BBC and Philips have combined to produce a two-disk set of interactive visual information about towns and villages all over the UK.

The Photodetector

The detector section of the pick-up has a two-fold purpose. Primarily, it has to detect the optical signal and convert this into an electrical pulse; but it also has to tell the signal processor if the beam is in or out of focus or is deviating from the spiral track. Again having decided that size, weight, cost and reliability are important, our most suitable choice of detector is a silicon photodiode. Photo-multipliers, though extremely sensitive, are bulky, require high voltages and are expensive; light-dependent resistors are insensitive and respond slowly to light; charge-coupled devices require complex processing. To cope with the weak signal which returns from the disk, which is the order of a few microwatts, a diode with high sensitivity is required.

We will discuss these detectors in Chapters 4 and 6.

This is the so-called *Foucault* or knife-edge focusing system.

Tracking and focusing signals are obtained by directing the double output beam from the wedge lens onto a pair of double-diode arrays (see Fig. 1.3). The tracking signal is generated by monitoring the difference in output between the total signal received by the two-diode arrays as shown in Fig. 1.5. When the laser is precisely on track an equivalent signal is obtained from both diode arrays. The difference between the two is zero. If the laser shifts either to the left or right, one pair of diodes will produce a greater signal than the other and an error voltage will be received at the output. This signal is fed to the signal processor and the laser is pulled back onto the track. Figure 1.5 also shows the means by which the focusing signal is generated. When the disk is in perfect focus, once again the signal from each diode segment will be equal. If the laser is either too close or too far away from the disk, uneven illumination of the segments of the diode array will result in an error signal, which again is used to pull the laser back into focus.

Encoding the Information on the Disk

Finally we have to consider how we encode our information about the real world, which is generally an analogue signal such as music or speech, into digital form on the disk. Thereafter, how do we reconvert the digital signal which the laser has read back into a form our human senses can cope with? To fully understand these processes takes us into the realms of signal processing and way beyond the scope of this text. We should, however, have a broad look at the techniques involved for the sake of seeing the entire CD system in context.

Other forms of information which can be stored on CD are computer data and video pictures.

For a full discussion of PCM and other forms of signal processing, see O'Reilly, J.J. *Telecommunications Principles* (Van Nostrand Reinhold, 1984), Chapter 6, or Wilson, J. and Hawkes, J. *Optoelectronics: An Introduction* (Prentice-Hall, 1983).

In some cases, recording is directly onto a digital tape recorder.

The process by which the analogue signal is converted to digital form uses a technique known as *pulse-code modulation* (PCM). The musical information is first recorded as a normal analogue signal on a master tape or disk. The recorded signal from each stereo channel is then monitored at regular intervals and the voltage at each sampling point converted into a 16-bit binary number, a process known as *quantization*. The rate at which the signal is sampled defines the highest frequency of signal which can be recorded. The optimum sampling frequency is determined by the so called *Nyquist criterion* which states that this sampling frequency should be twice that of the highest frequency of the signal to be recorded. The chosen sampling frequency of 44.1 kHz, one sampling point every 22.68 μs, sets an upper limit of about 22 kHz and ensures that the whole audio spectrum is encompassed. In principle, a lower sampling frequency of 36 kHz could be chosen since the audio spectrum only stretches to about 18 kHz. The higher rate is chosen to allow for the extended response of most modern amplifiers and also for the roll-off charac-teristic, which is inherent in any analogue filters used.

The audio spectrum ranges from about 50 Hz to about 18 kHz until your early 20's. As you get older the bandwidth decreases dramatically.

All analogue filters will pass a small portion of signals at those frequencies above their design cut-off.

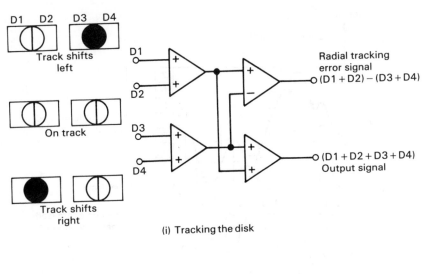

(i) Tracking the disk

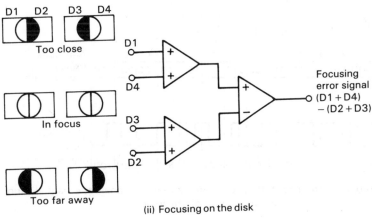

(ii) Focusing on the disk

Absence of light ◖   ◗ Presence of light

Fig. 1.5   Focusing and tracking of disk.

Figure 1.6 shows in simplified form the processes involved in converting an analogue signal into a digital logic signal. For the sake of clarity 3-bit quantization, where the signal is divided into $2^3 = 8$ levels, is shown as an example. In full 16-bit quantization, the analogue signal would be divided into 65 536 levels.

The actual process involved is considerably more complex than that shown above. Suffice to say here that the information from each stereo channel is quantized separately, then combined, a process known as *multiplexing*, and further encoded to incorporate error and tracking information, control and display information and synchronization data. The final digital signal is used to modulate the drive to a high-power laser which writes the signal onto the master disk by ablating the pits directly in the substrate material. The master disk is copied as many times as is required.

The process is known as 'eight-to-fourteen' modulation and is discussed in detail by Bowhuis, G. *et al., Principles of Optical Disk Systems* (Adam Hilger 1985).

A continuous-power argon or krypton laser is used here. These lasers will be discussed in Chapter 3.

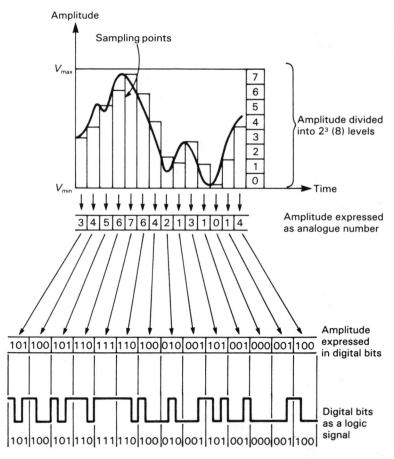

Fig. 1.6 Simplified representation of quantization of analogue signal and its conversion into digital logic.

It is the initial choice of 16-bit quantization which determines the inherently high dynamic range of the CD system as shown in the following example.

Worked Example 1.2   Calculate the dynamic range of an optical disk when 16-bit quantization is used to record the signal.

Solution: The use of 16-bit quantization means that at each sampling point the signal amplitude is converted into a 16-bit binary number. The loudest signal which can be recorded, therefore, corresponds to a binary number of 2^{16} and hence the dynamic range is $2^{16}:1$ or 65 536:1. Expressing this in decibels (dB) gives,

$$\text{Dynamic range} = 20 \log_{10} (2^{16})$$
$$= 96 \text{ dB}$$

Contrast the above figure with the 50 dB or so possible with an analogue system.

We should note here that the high dynamic range of the CD system is a function of the digitization process, and is not dependent on the fact that light is used to read the signal. Digital tape recorders can also show this performance, but the require-

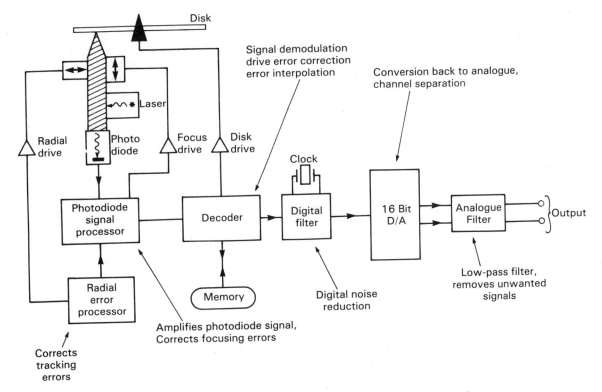

Fig. 1.7 Simplified representation of CD-player signal processing.

ment for fast access and the ability to store large amounts of data is likely to mean that the CD system will become the market standard. It should be borne in mind though that the dynamic range we have calculated is that of the *recorded* signal. The dynamic range of the signal which eventually reaches your ears will depend on, amongst other things, the linearity of the pre- and power amplifiers and on the power handling capacity of the loudspeakers.

(A personal opinion).

Decoding the Information on the Disk

Of equal importance to the laser in the CD player is the microprocessor-controlled decoding system. Here the information is extracted from the photodiode as a series of 0's and 1's, decoded and presented to the loudspeakers as an analogue signal. Figure 1.7 is a simplified block diagram of the entire CD player showing the optics and processing system. The digital signal from the photodiode is amplified and corrected for tracking and focusing errors before being demodulated. The signal is further checked for errors and control and display signals sent to the appropriate circuits. The left and right channels are separated and processed as separate frames of data entirely eliminating crosstalk. Decoding of the information differs in detail between the various machines on the market but all have the purpose of converting the digital signal back to analogue form. In essence, a digital filter removes any spurious signals introduced in decoding the quantized signal. This signal is then reconverted to analogue form before, finally, low-pass filtering removes any remaining unwanted signals and left- and right-channel signals are presented to the audio amplifier.

Crosstalk occurs when part of the signal from one channel is heard in the other.

A number of excellent publications are now available which describe the CD system in its entirety. For the definitive work written by the originators of the system see G. Bouwhuis, et al. *Principles of Optical Disk Systems* (Adam Hilger 1985). An excellent series of articles appeared in *Wireless World*, variously from January 1985 to 1986, under the general title of *Digital Audio* by J.R. Watkinson. Others include *The compact disk, player and system* by M. Sykes in *Electronics and Power*, May 1984 and *Digital audio is compact and rugged* by S. Miyaoka in *IEEE Spectrum*, March 1984.

Summary

In this first chapter, we have attempted to present the broad face of optoelectronics to the reader and show that optoelectronics provides the system designer with a radically new approach to design. By studying the design of the compact disk audio system we have seen a real example of how the use of optoelectronics frees the designer from the constraints of tradition and conventional technology, thereby enabling him to arrive at a system which is set to revolutionize audio replay.

Our outline of optoelectronic systems has brought out several important concepts as far as our approach in the rest of the book is concerned. At the heart of optoelectronics lies the use of light as our information carrying medium: an understanding of its nature is therefore crucial to our successful application of the techniques. This takes us nicely into the study of light sources and in particular the laser — its theory, operation and the types available. Following on from here we make a study of optical detectors and their properties, culminating in our bringing everything together with a look at some important optoelectronic systems and applications.

Problems

1.1 Calculate the approximate spot size to which a HeNe laser ($\lambda = 633$ nm) can be focused, using a lens of 0.45 NA.

1.2 Assuming the use of the lens and laser from the previous example, and making sensible assumptions about the packing of information on the disk, estimate the information density of an optical video disc of 12 ins diameter.

1.3 If an analogue signal is '4-bit quantized', how many levels of quantization does this represent?

1.4 In the range of CD players manufactured by Philips Ltd, the signal from the disc is read out at a frequency which is four times that at which the original signal is recorded. By consulting the references quoted in the text, decide what design advantages '4-times oversampling' provides.

1.5 The Philips marque of CD players use a 'single-beam' laser pick-up, whereas many other systems utilize a 'three-beam' pick-up. By consulting the references quoted in the text, compare and contrast the two systems.

Light and Laser Light

☐ To outline the behaviour and nature of light
☐ To discuss the processes of spontaneous and stimulated emission of photons
☐ To discuss the processes of amplification of light
☐ To describe the principles of laser action

We embark now on our discussion of the individual elements required in an opto-electronic system with a look at the first element in the chain: the light source. As indicated earlier, the invention of the laser is probably the single most important milestone in the development of optoelectronics. It is doubtful whether or not any of the other significant events, such as the progress made in optical fibres for example, would have occurred without the impetus and the promise of *great things to come* provided by the laser. Inevitably our treatment of light sources must concentrate on this remarkable device. We should not forget, however, the more traditional thermal sources such as arc sources, incandescent sources and gas discharges or even the simple light-emitting diode, which have an important, though less glamorous, role to play. We will refer to these sources where appropriate.

These sources are most often used for illumination. The most important for our purposes is probably the linear flashtube, which is widely used as a pumping source for pulsed lasers.

But firstly, our immediate task is to ask ourselves what light is and what its origins are, since this is fundamental to our study of optoelectronics itself. As stated in Chapter 1, light is the medium by which we carry information through our system. Only by having a firm grasp of its characteristics can we utilize this medium to its full extent and appreciate the unique properties of the laser. In this chapter, we will concentrate on a study of the nature of light and the physical processes leading to laser action in an atomic system; in the next, we will look at laser systems themselves.

The Behaviour of Light

Light is not the easiest of natural phenomena to describe. For many centuries scientists have debated, and argued over, the nature of light. We will not take part in this debate but will try to present a description of light which will satisfy our working needs. Do not be disheartened to realise that we have difficulty in discussing such a fundamental phenomenon. The problem is not with light itself but with our inability to model its complex character under one mathematical theory. In fact to completely describe the properties of light requires us to adopt two different models of behaviour: the *electromagnetic-wave* model and the *photon* model.

The Wave Model of Light

When describing optical phenomena like diffraction, interference and polarization, it is convenient to model the behaviour of light on that of a travelling wave

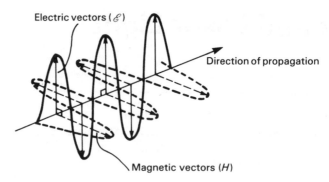

Electric vectors ($\mathscr{E}$)

Direction of propagation

Magnetic vectors (H)

Fig. 2.1 Field representation of an electromagnetic wave showing relative orientation of $\mathscr{E}$ and H vectors. The wave shown is linearly polarized in the vertical plane of the $\mathscr{E}$ vector. In general the vectors will be randomly orientated in time about the axis.

For detailed discussion of electromagnetic theory, see texts such as Compton, A.J. *Basic Electromagnetism*, (Van Nostrand Reinhold, 1986), Carter, R.G. *Electromagnetism for Electronic Engineers* (Van Nostrand Reinhold, 1986) or Born, M. and Wolfe *Principles of Optics* (Pergamon).

Electromagnetic theory was first proposed in mathematical form by the Scottish Scientist James Clerk Maxwell in the late 19th century.

propagating through space. The wave can be described in terms of a combination of mutually perpendicular electric and magnetic fields, known as an *electromagnetic wave*. The direction of propagation of the wave is at right angles to both field directions (Fig. 2.1)

The theory of electromagnetic radiation is one of the corner-stones of electrical engineering and embraces such phenomena as light, radio waves, microwaves, X-rays and gamma rays which differ only in their characteristic wavelength. Visible light, however, stretching from a wavelength of about 400 nm at the violet end to around 700 nm at the red end, forms only a small part of that continuum known as the *electromagnetic spectrum* (Fig. 2.2). In optoelectronics, though, we usually treat light as stretching out to 200 nm, the far ultra-violet, at the short-wavelength end and out to about 15 μm, the far infra-red, at the long-wavelength end.

A complete, formalized description of the propagation of light through free space and the energy transferred by it, requires that we treat electromagnetic radiation as a *vector* wave and thus specify it, in 3-dimensional co-ordinates, in terms of its magnitude and direction. For most purposes, though, we may represent a travelling light wave as a one-dimensional scalar wave, as shown in Fig. 2.3, provided that we remember it has a direction of propagation. We need only describe such a wave in terms of either the electric field or the magnetic field: both are not necessary, since we can always extract one from the other. Conventionally we use the amplitude of the electric-field vector $\mathscr{E}$ to describe a plane wave of angular frequency ω and wave vector k at any point in time and space, by,

$$\mathscr{E} = \mathscr{E}_0 \sin(\omega t - kx - \phi) \tag{2.1}$$

where $\mathscr{E}_0$ is the maximum amplitude of the electric vector; ϕ is a phase constant which is independent of position and time and depends on the chosen location of the origin; t and x are the respective time and space coordinates.

In the above relation the angular frequency of the wave is related to its linear frequency f by

$$\omega = 2\pi f$$

and the magnitude of the wave vector is given by

$$k = 2\pi/\lambda$$

where λ is the wavelength.

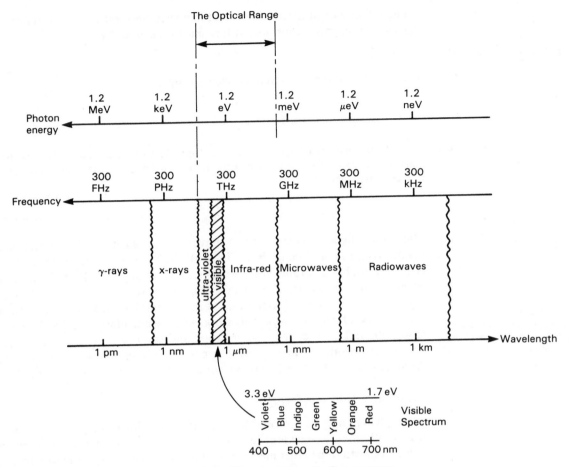

Fig. 2.2   The electromagnetic spectrum.

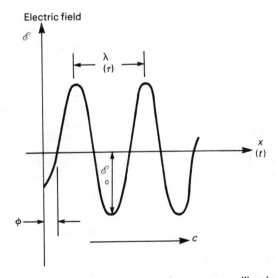

Fig. 2.3   A one-dimensional electromagnetic wave travelling in the x direction.

The velocity of propagation of light in a vacuum is related to the fundamental electric and magnetic constants of free space, ϵ_0 and μ_0, by

$\epsilon_0 = 8.87 \times 10^{-12}$ F m^{-1} and $\mu_0 = 4\pi \times 10^{-7}$ H m^{-1}.

$$c = (\epsilon_0 \mu_0)^{-\frac{1}{2}}$$

(2.2)

and also, to its frequency and wavelength, by

$$c = f\lambda$$

(2.3)

Exercise 2.1 Calculate the velocity of light in a vacuum from the free-space values of the electric and magnetic constants. Consider also how the model could be changed to account for propagation in transparent media.

[299.5 × 10⁶ m s⁻¹]

The beam of light we have described above oscillates such that the orientation of its electric vector is random in time (Fig. 2.4). Such a wave is *unpolarized*. A beam, however, can be constrained to vibrate in preferential planes of oscillation: such a beam is said to be *polarized*. When the light wave oscillates in a single plane of the electric vector it is *linearly polarized*. A more complex situation occurs when the plane of polarization continually rotates perpendicularly to the direction of propagation. This is *circular* polarization and can occur in two senses: *left* and *right* polarization. The most general case is that of *elliptical* polarization in which the amplitude of the polarization vector also changes with the rotation.

Certain materials have the ability to alter the mode of polarization of a beam of light passing through it. More of this in later chapters.

The Photon Model of Light

Although light may be described in terms of a travelling wave, in some instances it helps to discuss its behaviour in terms of the amount of energy it imparts in an interaction with some other medium. In this case we can imagine a beam of light to be composed of a stream of small lumps or *quanta* of energy, known as *photons*.

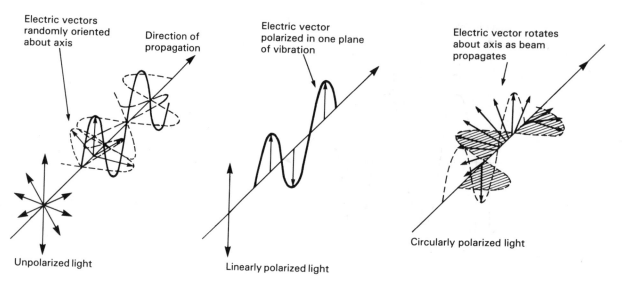

Electric vectors randomly oriented about axis

Direction of propagation

Electric vector polarized in one plane of vibration

Electric vector rotates about axis as beam propagates

Unpolarized light

Linearly polarized light

Circularly polarized light

Fig. 2.4   Polarization of light.

Each photon carries with it a precisely defined amount of energy which depends only on its wavelength. Even although a photon can be thought of as a particle of energy, it still has a fundamental wavelength associated with it which is equivalent to that of the propagating wave. The energy of a single photon is given in terms of its frequency f or wavelength λ, as

$$W_{ph} = hf = hc/\lambda \qquad (2.4)$$

where h is Planck's constant and c is the velocity of propagation of the photon in free space. How the energy and frequency of a photon are linked are shown in eqn (2.4); knowing one gives us the other. The photon description of light is useful when the light source is weak, in other words, when we are dealing with relatively few of them.

The concept of a photon is principally due to the work of three men: Max Planck, Neils Bohr and Albert Einstein, in the period from around 1900 to 1920. By invoking the idea of light being emitted in tiny pulses of energy, they were able to explain such phenomena as black-body radiation, emission of light from atoms and the photoelectric effect.

Planck's constant $h = 663 \times 10^{-36}$ J s.

To illustrate the use of eqn (2.4) we can calculate the energy of a photon of 600 nm wavelength. This photon would be in the red part of the spectrum. The energy of a single photon is,

$$
\begin{aligned}
W_{ph} &= hc/\lambda \\
&= 663 \times 10^{-36} \text{ J s} \times 300 \times 10^6 \text{ m s}^{-1}/600 \times 10^{-9} \text{ m} \\
&= 332 \times 10^{-21} \text{ J}
\end{aligned}
$$

Worked Example 2.1

This is a tiny amount of energy! We can put this into more manageable numbers by using the concept of the electron volt. An electron volt (eV) is defined as the energy obtained by an electron when it is accelerated through a potential difference of 1 volt. Hence,

$$1 \text{ eV} \equiv 160 \times 10^{-21} \text{ J}$$

$W_{electron} = q_e V = 160 \times 10^{-21}$ C $\times 1$ V $= 160 \times 10^{-21}$ J.

Therefore,

$$
\begin{aligned}
W_{ph} &= 332 \times 10^{-21} \text{ J}/160 \times 10^{-21} \text{ J eV}^{-1} \\
&= 2.07 \text{ eV}
\end{aligned}
$$

Visible photons range in energy from 1.74 eV (700 nm) to 3.34 eV (400 nm).

The Origins of Light

To understand the operation of the laser and other light sources we need to appreciate the unique character of the light emitted from gases and solids. All radiating bodies when viewed by the naked eye appear to possess a characteristic colour. Sunlight is white, a piece of hot iron may be orange-red and a sodium street lamp is yellow. If light, however, from any of these sources is passed through a prism, it spreads out in a series of component colours known as a *spectrum*. Sunlight appears as a continuous band of colours ranging from red through to violet, the piece of iron also shows a continuum from dull red to orange and the sodium lamp displays a series of bright, narrow lines. Whether the spectral distribution is continuous or is in discrete lines depends on the nature and temperature of the source.

Exercise 2.2 The human eye perceives colours using three sets of sensors tuned to wide bands of red, green and blue light. The sodium lamp mentioned above, composed of primarily a single monochromatic line, stimulates the red and green equally: we see this as 'yellow'. A yellow line on a television screen, however, is produced using two separate red and green luminous phosphors. Although this emission profile is distinctly 'double-humped' and not monochromatic, the eye still sees this as 'yellow'. Why then are there no 'purple' lasers?

A traditional description of the quantum model is given by Anderson, J.C., Leaver, K.D. Rawlings, R.D. and Alexander, J.M. *Materials Science* (Van Nostrand Reinhold, 1985) Chapters 1–5. For a highly readable and lucid account of the historical development of our knowledge of atomic structure I thoroughly recommend Hoffman, B. *The Strange Story of the Quantum* (Penguin, 1982).

This is the force which acts along a string when a stone is swung around your head.

This simple treatment ignores divalent atoms and X-ray emission.

We will return to this concept of spontaneous de-excitation, and the alternative process of stimulated de-excitation, when we discuss the laser later in this chapter.

Spontaneous Emission of Photons

The origins of spectral emission of light from a material can be traced back to the existence of allowed levels of energy within the structure of the atom itself and successive excitation and de-excitation of electrons between these levels. On the basis of the quantum model of the atom, electrons are not permitted a continuum of orbits around the nucleus but are confined to a set of discrete stable orbits by a balance between two opposing forces. The force of electrostatic attraction, known as the Coulomb electrostatic force, pulls the negative electron towards the positive nucleus, whereas the force due to the mass acceleration of the electron acts radially outwards from the nucleus. Bound electrons cannot take up any value of energy, but can only possess specific discrete energies according to the allowed orbits. The energy is said to be *quantized* and the permitted values of energy are known as *energy levels*. The further the electron is from the nucleus the less tightly bound it will be.

For the simplest case of a hydrogen atom, which has only one electron, the range of energies which the electron can possess are shown in the *energy level diagram* of Fig. 2.5. With the electron in its lowest possible energy level, the *atom* is said to be in its *ground state*. An atom can only be raised into a higher *excited state* if the electron *absorbs* an amount of energy exactly equal to the difference between the lower and higher levels. For example, to raise the hydrogen atom from a lower state W_l to an excited state W_u requires that the absorbed energy, ΔW, equals $W_u - W_l$. The energy may be supplied as, for example, light, heat or collisional energy. What is important is that it must be of the required amount. For atoms other than hydrogen, with more than one electron, the allowed energy levels fill with electrons from the lowest energy up, according to the *Pauli Exclusion Principle*. The ground state of a many-electron atom corresponds to the situation with all available electrons packed into the lowest possible levels. Generally, excitation of the atom raises the outer orbital electrons, the so-called valence electrons, into higher levels.

When in an excited state, the atom may be further excited into an even higher state or it may release some, or all, of its excess energy by dropping back to a lower state. Generally, the lifetime of an atom in an excited state is around 10 ns before it *spontaneously* relaxes to a lower level with the release of a quantum of energy equal to the difference between its initial excited state and its final lower state. Although this energy may manifest itself in several different forms, our interest is in those energy transitions, known as *optical transitions*, which result in the emission of a photon of light.

The energy of the emitted photon W_{ph} depends only on the difference between the two energy levels, and is given by, for the two levels considered earlier,

$$W_{ph} = W_u - W_l = \Delta W_{ul} \tag{2.5}$$

From eqn (2.4), we may remember that the photon energy is also given by,

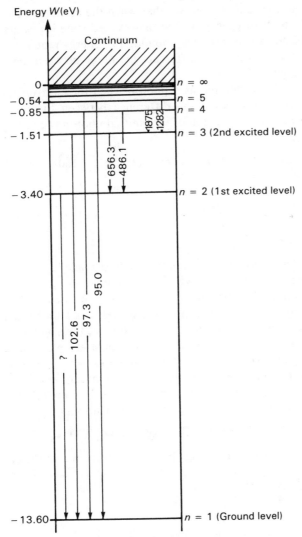

Fig. 2.5 Energy diagram for hydrogen showing principal spectral lines (all wavelengths given in nanometers).

$$W_{ph} = hf_{ul} = hc/\lambda_{ul}$$

where f_{ul} is the frequency of the photon emitted and λ_{ul} is its wavelength. The
relation of frequency and wavelength of a photon to the energy difference between
its initial and final states are given in eqns (2.4) and (2.5).

Because there are a number of possible energy levels available to the electron, a
range of optical transitions is possible, giving rise to the characteristic spectrum for
that element. Some of the possible transitions for hydrogen are shown in Fig. 2.5.

How much energy does a hydrogen atom need to absorb to be raised from its lowest
(ground) state to its first excited state? Also, what is the wavelength of the photon
emitted when the atom drops back to its ground state?

Worked Example 2.2

Solution: From Fig. 2.5, we can see that when the atom is in its ground state it possesses -13.6 eV of energy. In the first excited state it possesses -3.4 eV of energy. The electron is less strongly bound. The energy required to lift the atom into the higher level is just the difference in energy between the two. Hence the absorption energy is 10.2 eV.

The energy which is emitted as a photon when the electron falls back to the ground state must also equal the difference between the two levels, that is -10.2 eV. Hence the wavelength of the emitted photon is given by,

$$\lambda = hc/W_{ph}$$
$$= 663 \times 10^{-36} \text{ J s} \times 300 \times 10^6 \text{ m s}^1/(10.2 \text{ eV} \times 160 \times 10^{-21} \text{ J eV}^{-1})$$
$$= 121 \times 10^{-9} \text{ m}$$
$$\lambda = 121 \text{ nm}$$

Population Densities

The situation described above relates to that of a single isolated atom. Consider now the more realistic situation where vast numbers of atoms exist in different states of excitation. In this case the electrons associated with each atom are distributed across the whole range of energy levels available to them. At thermal equilibrium the numbers of atoms per unit volume, the *population density*, in a particular energy state depends solely on the temperature of the gas and the difference in energy between the excited level and a known lower level.

For a gas in thermal equilibrium with its surroundings at a temperature T, the population density of atoms N_u in an excited state W_u in relation to those N_l in a lower energy state W_l is given by the Boltzmann relation, as

$$N_u/N_l = \exp[-(W_u - W_l)/kT] = \exp[-\Delta W_{ul}/kT] \tag{2.7}$$

and shown diagrammatically in Fig. 2.6. The significance of the above relation is

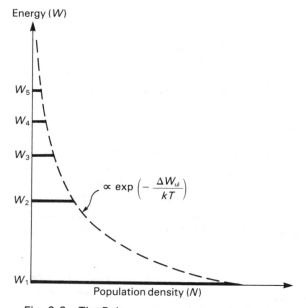

Fig. 2.6   The Boltzmann population distribution.

that at normal temperatures the vast majority of atoms will be in their ground states. Relatively few of them will be in higher states of excitation. Raising the temperature of the system pushes more atoms into the higher levels of excitation which subsequently de-excite by spontaneous release of photons back to the lower levels. A dynamic situation exists in which the population ratio of any pair of states at a particular temperature is governed by the Boltzmann equation at that temperature.

We will see later that this behaviour is in contrast to that existing within a laser medium, where we have a large population density in an upper state: a non-equilibrium population inversion.

Thermal Light Sources

The processes we have just described form the basis of light emission from thermal sources. If a low-pressure gas such as hydrogen is enclosed in an evacuated glass tube and excited by the passage of an electric current through it, the constituent atoms and molecules continually absorb energy by collisions with electrons and also with each other. No sooner do the atoms gain energy than they lose it again by a mixture of spontaneous emission of photons, heat and collisional de-excitation. In such a dynamic situation the energy level populations are governed by the Boltzmann relation at the specific temperature of the gas. Because the atoms and molecules are sufficiently far apart that the energy structure pertains to that of an isolated atom, photon emission will cover a wide range of wavelengths but be discrete in nature. Such sources of light are known as *gas discharge sources* and are widely used as reference sources in spectroscopy. Typical examples are the sodium-vapour street lamps mentioned earlier and the mercury-vapour lamp. A spectrum for the mercury-vapour lamp is shown in Fig. 2.7.

When atoms are closer together, as in a dense gas like the sun or a solid such as a piece of hot iron, the energy structures associated with individual atoms influence one another. This is the collective behaviour of many interacting atoms of different species, rather than the characteristic behaviour of individual atoms of a particular element. The isolated levels must shift slightly in energy and overlap in relation to each other to accommodate all the available electrons. Instead of individual, discrete energy levels for the isolated atom, we find that *bands* of closely spaced energy levels develop. In between these bands are *forbidden zones* of energy,

On the atomic scale, 'far apart' means that the atoms are separated by distances of several atomic diameters.

For details of such lamps see Levi, L. *Applied Optics* (Wiley, 1968) Vol. 1.

This behaviour can be explained on the basis of the *Pauli Exclusion Principle*, see Anderson, J.C., Leaver, K.D., Rawlings, R.D. and Alexander, J.M. *Materials Science* (Van Nostrand Reinhold, 1985).

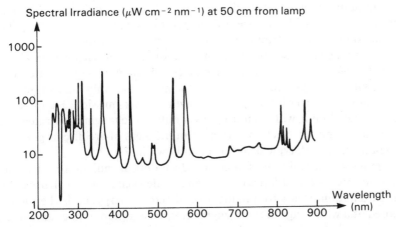

Fig. 2.7   Spectrum of 1000 W mercury (arc) lamp. (Adapted from data supplied by ORIEL Ltd.)

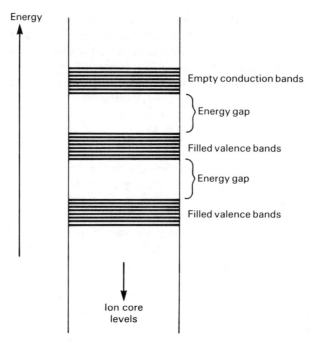

Fig. 2.8　Energy bands in a solid.

I have compressed a lot of 'heavy' physics here into a few paragraphs. Again I refer you to a text like Anderson's *Materials Science* for a more leisurely and explicit discussion of these points.

energy gaps, which atoms cannot occupy (Fig. 2.8). As for a gas at a given temperature, the lowest available energy levels are filled with electrons: these are known as the *valence bands*. Above the valence bands are a range of primarily empty bands, these are the *conduction bands*.

Excitation raises atoms from the heavily populated valence bands into the relatively empty conduction bands. Emission of a photon occurs when an electron in a conduction band relaxes to a valence band with the release of a corresponding amount of energy. Because emission occurs between *bands* of energy, rather than the well-defined levels described earlier, the characteristic spectrum is a broadband continuum in which the individual transitions are indistinct. Although the relative populations of bands are still temperature dependent, they are governed, not by the Boltzmann distribution, but by *Fermi-Dirac* statistics.

The concept of black-body radiation is fundamental to optoelectronics. Because of this the main points are outlined in Appendix A.

Again reference should be made to Levi, L. *Applied Optics* (Wiley) Vol. 1 for details of such lamps.

The overall spectral distribution of energy radiated from such a thermal source is given by the *black-body* relation which gives the radiant energy emitted by a hot body at a known temperature. In essence the hotter the body is the more energy it will emit and the shorter will be its wavelength of peak emission. Hence, as a piece of iron is heated up it will begin to glow a dull red, changing to orange-red and yellow as it gets hotter. Such materials form the basis of arc discharge sources or incandescent lamps. A typical spectrum is shown in Fig. 2.9.

In summary, thermal light sources produce optical radiation as a result of a dynamic process of excitation and spontaneous de-excitation of valence electrons. Because excitation occurs to a number of different energy levels the radiation is composed of a wide range of wavelengths emitted at random times and in random directions: the light is said to be *incoherent*. This behaviour is in stark contrast to that pertaining to a laser in which the light emitted is almost monochromatic, in

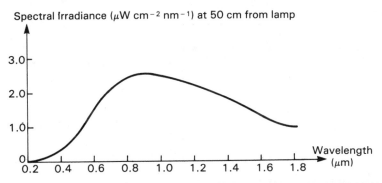

Spectral Irradiance (μW cm^{-2} nm^{-1}) at 50 cm from lamp

Fig. 2.9   Spectrum of 100 W tungsten-halogen lamp operating at 3200 K. (Adapted from data supplied by ORIEL Ltd.)

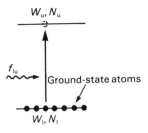

Stimulated absorption

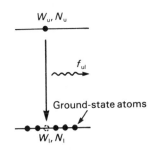

Spontaneous emission

Stimulated emission was first proposed by Albert Einstein in 1913 as a mechanism of stellar light emission and is believed to occur naturally at the sun's core.

phase and unidirectional: this light is said to be *coherent*. Just why this is so we will see in the remainder of this chapter and the next.

Amplification of Light

For atomic systems in thermal equilibrium with their surroundings, emission of light is the result of two main processes, namely, *absorption* of energy and, subsequently, *spontaneous emission* of energy. A third mechanism exists, however, whereby an atom in an upper energy level can be triggered or *stimulated* in phase with an incoming photon to emit a photon of the same energy as that of the stimulating photon. This is the process known as *stimulated emission* which, although not a dominant process in thermal systems at room temperatures, is crucial to the formation of laser action.

Population Inversion

The key to the above behaviour lies in the likelihood, or *probability*, that an optical transition will actually occur. What we have omitted to say up until now is that not all transitions, upwards or downwards, occur with equal probability: some are more likely than others to absorb or emit a photon. For spontaneous emission of a photon, the probability of occurence is inversely related to the average length of time t_{spont} that an atom can reside in the upper level of the transition before it spontaneously relaxes. Typically t_{spont} is some tens of nanoseconds. The shorter the spontaneous lifetime, the greater is the likelihood that spontaneous emission will occur.

Defined by the Einstein *A* coefficient. See Appendix B.

For some pairs of levels in certain materials the spontaneous lifetime can be the order of a few milliseconds and hence the likelihood that a spontaneous transition will take place is relatively low. As the likelihood of spontaneous emission decreases, however, the conditions which favour stimulated emission are enhanced. Because of the existence of these longer-lifetime states, it is possible to create a situation where the rate at which atoms are *pumped* into one of these states exceeds the rate at which they leave. Hence, a large number of atoms can be excited into, and held in, the upper state leaving an almost empty state below them. Such a situation known as *population inversion* is obviously not covered by the Boltzmann relation and the system is no longer in thermal equilibrium.

The Einstein *B* coefficient, see Appendix B.

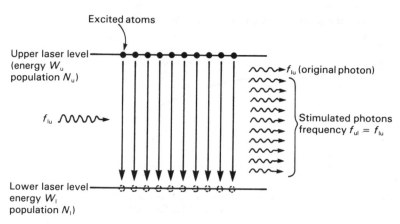

Fig. 2.10 Origin of stimulated emission of photons in an atomic system where $N_u \gg N_l$.

Laser Emission

See later in this chapter for an expansion of these points.

Irradiating an atomic system, in which a population inversion has been created, with a beam of photons of frequency f_{ul} corresponding to the transition between the upper and lower states, vibrates the atoms in the upper state in phase with the incident light and so stimulates the emission of a cascade of photons each of frequency f_{ul} (Fig. 2.10). The emitted photons all possess the same wavelength and vibrate in phase with the incident photons. What we have done is to add photons to the incoming beam by promoting stimulated emission at the expense of spontaneous emission: *we have amplified light!*

In reality, as we will see in the next few pages, it is light *oscillation* that we are actually dealing with. Perhaps the laser should be called a LOSER?

The above behaviour forms the basis of laser action and gives us the origin of the name LASER: it is an acronym for Light Amplification by Stimulated Emission of Radiation.

Three- and Four-Level Laser Systems

See Appendix B.

In practice it is not possible to create a working laser based on absorption and emission between only two energy levels as described above. Because as it happens, for any pair of levels, the rate at which the upper level is populated by absorption equals that at which atoms leave by stimulated emission, the best we can hope for in a two-level system is an equality of populations in the upper and lower levels. Population inversion cannot be achieved. Systems involving a three- or even four-level structure are necessary.

In a three-level system, atoms are pumped into the highest level with de-excitation occurring to the long-lifetime intermediate level which serves as the upper level of the laser transition. Consider the three-level atomic structure consisting of levels p, u and l, as shown in Fig. 2.11. The energy of the upper pumping level is denoted by W_p, that of the upper level of the laser transition by W_u, and the lower (ground) level by W_l. Initially if the system is in thermal equilibrium with its surroundings then the populations N_p, N_u and N_l, of the respective levels are given in accordance with the Boltzmann relation, and thus N_p is much greater than N_u which is, in turn, much greater than N_l.

This energy may be supplied in one of many forms. See Chapter 3.

Supplying pump energy to the system, equivalent in magnitude to W_p, will raise atoms in the ground state l to the upper pump level p. De-excitation from the pump

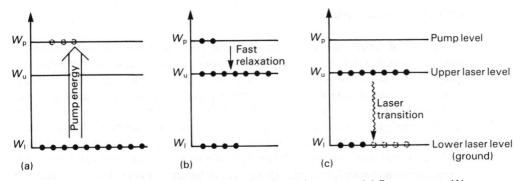

Fig. 2.11 Process of laser action in a three-level atomic system. (a) Pump energy W_p populates level p from level 1. (b) Rapid relaxation from p→u establishes population inversion between levels u and l. (c) Spontaneous photon stimulates cascade of emission in laser transition (u→l).

level to the upper laser level can occur by spontaneous emission or by non-radiative processes such as collisional de-excitation. If the rate at which the upper laser level is fed from the pump level exceeds that of spontaneous de-excitation from the laser level, then a population inversion can be established between u and l, such that N_u becomes much greater than N_l. Ideally, then, the rate at which the upper laser level is fed from the pumping level should be rapid with consequent depopulation of the ground level taking place relatively slowly. The magnitude of the population inversion is simply,

$$N_{inv} = N_u - N_l \qquad (2.8)$$

The system is now primed for laser action. Spontaneous emission of a photon between the laser levels is enough to stimulate an avalanche of coherent photons which are characteristic of laser action.

An improvement on this behaviour is obtained with a four-level structure, where the laser transition takes place between the third and second excited states (Fig. 2.12). In this case we want depopulation of the lower laser level to be rapid to ensure that the upper level is always full and the lower level always empty.

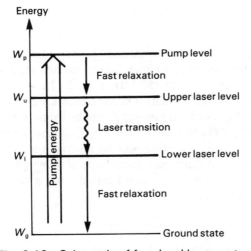

Fig. 2.12 Schematic of four-level laser system.

Optical Resonant Cavities

We have shown then that to obtain light amplification from a given medium we need to establish a population inversion between a chosen pair of energy levels and promote stimulated emission at the expense of spontaneous emission. By itself, however, the above statement is not a sufficient criterion to produce laser action in a given system. If we do not confine this system in a special way, it would radiate spontaneously in so many different directions that we would not be able to sustain stimulated emission. In practice we need to provide some means of optical feedback such that the stimulated beam is made to pass backwards and forwards several times allowing it to stimulate further emission as it goes. In other words we want the overall gain of the system to be positive. This is achieved by bounding the laser medium between two mirrors, one totally reflecting and the other partially reflecting. The reflectivity of the partial reflector is normally between 10 to 90%

We will return to this point later in this chapter.

and is necessary to ensure that some laser light can escape and provide useful optical power.

Longitudinal Cavity Modes

By bounding the laser medium between a pair of mirrors and allowing light to bounce back and forth, we have created an optical *resonant cavity* in which only specific *modes of oscillation* can be supported. An oscillation mode in a resonant cavity can be likened to the establishment of standing waves on a stretched string pinned at both ends: only those modes corresponding to multiples of *half a wavelength* can be supported, all other modes will die away. This condition is expressed as,

$$\tfrac{1}{2}\bar{n}\lambda = L$$

The principles of resonant cavities are fundamental to a study of e.m. radiation. For further information, see: Compton, A.J., *Basic Electromagnetism* (Van Nostrand Reinhold, 1986); Carter, R.G., *Electromagnetism for Electronic Engineers* (Van Nostrand Reinhold, 1986); Bleaney, B. and Bleaney, B.I., *Electricity and Magnetism* (Oxford University Press, 1976). For special reference to optical cavities, see Yariv, A., *Optical Electronics* (Holt, Reinhart and Winston, 1984) or Wilson, J. and Hawkes, J.F.B., *Optoelectronics* (Prentice-Hall, 1983).

where L is the cavity length and $\bar{n}$ is the number of modes which can be supported.

The frequency of each mode is given by

$$f_m = \bar{n}c/2L \tag{2.9}$$

and the spacing between modes given by

$$\frac{\Delta f_m}{\Delta \bar{n}} = \frac{c}{2L} \tag{2.10}$$

Putting $\bar{n} = 1$ in eqn (2.10) gives the spacing between adjacent modes. For a laser cavity, which may be a metre or so long, several-million modes of oscillation are possible, each separated by about 150 MHz (Fig. 2.13).

We appear now to have an anomaly. On one hand we are saying that only certain modes which are multiples of a frequency f_m can be supported within a cavity, yet on the other, only those photons with frequency f_{ul} can be amplified. Do f_m and f_{ul}

(It would have to be or we would never have any lasers!)

coincide? If we treat f_{ul} as an infinitesimally narrow line then this would seem unlikely. In fact the actual situation is more hopeful than that because the spectral line emitted by the transition from u to l actually has a *finite* linewidth Δf_{ul} as described in the next section.

Spectral Linewidths

We have assumed up to now that the spectral width of the emitted laser line is

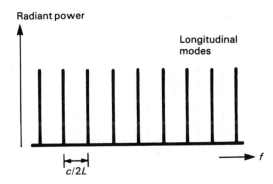

Fig. 2.13   Some possible oscillation modes within a longitudinal cavity.

infinitesimally narrow with a frequency of exactly $f_{ul} = \Delta W_{ul}/h$. This in fact is not the case. The emitted spectral line will have a *finite* linewidth Δf_{ul} centred around f_{ul} (Fig. 2.14). The width of this profile relates to the purity or *monochromaticity* of the light. The narrower the line the more monochromatic is the laser light, that is, the fewer frequencies it is composed of. Several mechanisms can occur which result in this broadening of the spectral line. It is not necessary for us to go into these mechanisms; it will suffice for our purpose to appreciate that because of their influence, the linewidth can be as much as a few nanometres. Actual values of the linewidth depend on the laser medium and optical configuration and will be outlined when we discuss the various types of laser.

'Monochromatic' means composed of one colour or, more specifically, of one frequency.

For a discussion of line broadening mechanisms see Yariv, A. *Optical Electronics* (Holt Rinehart Winston 1985).

Generally, linewidth is quoted as Δf (in Hz) or as $\Delta \lambda$ (in m). To convert from one to the other, say Δf to $\Delta \lambda$, we have

$$f = c/\lambda$$

thus

$$\Delta f/\Delta \lambda = c/|\lambda|^2$$

therefore,

$$\Delta f = c\,\Delta \lambda/\lambda^2 \tag{2.11}$$

Returning to our optical cavity, we can now see that the only modes of oscillation which can be supported within the cavity are those which coincide with the line

To add to the confusion, the linewidth is sometimes quoted in terms of the *wavenumber*, defined as $1/\lambda = f/100c = W/100hc$ (in cm^{-1}). Thus $\Delta(1/\lambda) = f/100c$.

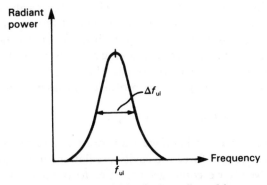

Fig. 2.14   Spectral linewidth of a laser line of frequency f_{ul}.

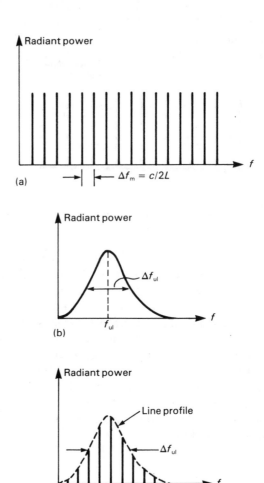

Fig. 2.15 Longitudinal oscillation modes within a laser cavity. (a) Possible cavity modes. (b) Laser line profile. (c) Allowed cavity modes.

profile of the laser transition (Fig. 2.15). Energy will build up in these allowed modes until such times as the gain of the system exceeds the accumulative effect of the losses. In other words laser action takes place when the gain provided by stimulated emission from the upper laser level just exceeds any losses due to, for example, spontaneous emission, off-axis photons and absorption and scattering at the mirrors. Laser light is emitted in a highly directional beam along the optic axis of the cavity. The laser line we are now talking about is not a single infinitesimally narrow line but is composed of a series of a few lines separated by a fixed frequency whose intensity is governed by the line profile of the laser transition.

Single-Frequency Operation

We should now be able to see that if we incorporate some element of *fine tuning* of the cavity length we could force the cavity oscillations into a *single longitudinal mode*. In other words we could get the laser to oscillate at one single cavity frequency.

We will see the importance of this when we come to discuss some laser applications in Chapter 6.

28

In practice this is achieved by inserting a frequency selective component, for example an etalon, into the cavity. An etalon may be nothing more than a block of quartz of a specific thickness and with parallel sides. By tilting the etalon in the cavity we finely alter the length of the cavity such that only one mode satisfies the cavity-length criterion. The unwanted modes are lost. To ensure that we get the most available power from the laser we try to select the frequency mode nearest to the centre of the line profile. There is of course a power penalty here. By going from a single wavelength, multi-frequency operation to single frequency we lose about half of the possible radiant power.

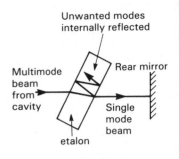

Practical etalons are usually composed of two blocks of quartz separated by a thin air space and enclosed in a temperature-controlled oven to ensure that the laser stays in the selected mode.

Transverse Cavity Modes

In the above discussion, we have tacitly ignored any width to the cavity; the modes we have been talking about are the *longitudinal* or *axial* modes. Of course the cavity will have a finite width and will support *transverse* modes arising from waves travelling off-axis along the cavity. These modes influence the *spatial* profile of the beam. Such modes are defined in terms of the Transverse Electro Magnetic wave distribution across the cavity, *TEM* modes. The *fundamental* mode is the TEM_{00} mode and corresponds to a smooth distribution of light across the output of the laser. This and some higher-order modes, such as TEM_{01} or TEM_{11}, are shown in Fig. 2.16. The transverse modes are a function of the cavity width. By placing a variable aperture in the cavity the diameter can be reduced to the point where a single TEM_{00} mode is sustained. Again we lose radiant flux in going from multi- to single-mode.

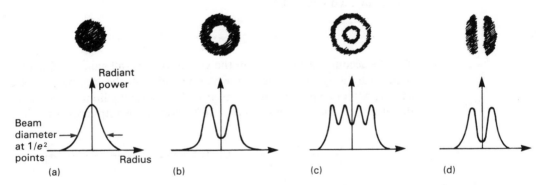

Fig. 2.16 Some typical transverse cavity modes. (a) Fundamental TEM_{00} mode (Gaussian). (b) First-order TEM^*_{01} mode (doughnut). (c) Higher-order mode. (d) First-order TEM_{01} mode.

Threshold Condition for Laser Action

In analysing the conditions required to produce laser oscillation in a given system, we have not accounted for any power losses which may occur. Although a population inversion is essential for laser action to proceed, no useful laser light will be produced unless the radiant power produced by stimulated emission exceeds that lost by other mechanisms. System losses include power emitted in spontaneous transitions, internal losses such as absorption in the medium in unwanted transitions, scattering in the medium and scattering and diffraction at the mirrors.

These losses represent wasted energy. Transmission of laser light through the output mirror is also a power loss but is an essential condition for the extraction of useful power from the laser. A threshold must therefore exist, before laser action can occur, corresponding to the point of balance between gain and loss.

The Threshold Gain of the System

We can estimate the threshold condition for a given system by examining Fig. 2.17. The laser medium of length L is enclosed at each end by two mirrors, one perfectly reflecting and the other with a reflectivity R. Initially we can imagine a beam of radiant power Φ_0 leaving the total reflector and traversing the active medium. The fractional change in power per unit length L of the beam's path can be defined in terms of a *small-signal gain coefficient* g of the medium as,

$$g \equiv \frac{1}{\Phi} \frac{d\Phi}{dL} \tag{2.12}$$

SI units of g are m^{-1}.

which after integration yields,

Such exponential relationships are common in naturally occurring phenomena, for example radioactive decay.

$$\Phi = \Phi_0 \exp gL \tag{2.13}$$

For amplification of light it is necessary that the gain coefficient of the system be positive. If g is negative, which is the situation pertaining to thermal equilibrium, the beam will lose energy as it traverses the medium.

Exercise 2.3   Calculate the gain coefficient g of a laser system in which the power of an incident light beam doubles over a distance of 1 m.

[0.69]

To take account of the losses in the system the overall gain coefficient g must be reduced by a loss coefficient g_i which includes all the internal losses described earlier. After one pass of the beam through the laser medium its radiant power will therefore become $\Phi_0 \exp[(g - g_i)L]$. Reflection at the output mirror will reduce the

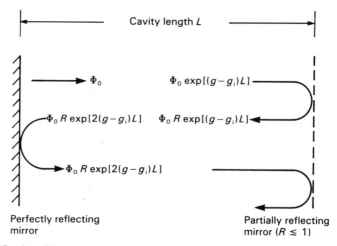

Fig. 2.17   Amplification of light in a laser medium of gain g and internal losses g_i.

power to $\Phi_0 R\exp[(g - g_i)L]$. Another pass through the medium yields a power of $\Phi_0 R\exp[2(g - g_i)L]$. Finally, after reflection at the total reflector, the beam will have made a complete circuit through the system and its radiant power will now be given as $\Phi_0 R\exp[2(g - g_i)L]$. If at this stage the new power exceeds its initial value, laser oscillations can be sustained. Hence the minimum condition for oscillations to proceed is that,

$$\Phi_0 R\exp[2(g - g_i)L] = \Phi_0$$

or

$$R\exp[2(g - g_i)L] = 1 \qquad (2.14)$$

Rearranging gives

$$\exp[2(g - g_i)L] = 1/R$$

Taking natural logarithms of both sides, we have

$$2(g - g_i)L = \ln(1/R)$$

and we arrive at the following condition for the small-signal gain coefficient,

$$g = g_i + (1/2L)\ln(1/R) \qquad (2.15)$$

Since R is usually very close to unity,

$$\ln(1/R) \simeq 1 - R = t$$

where t is the transmission of the output mirror.

We now have

$$g_{th} \simeq g_i + t/2L \qquad (2.16)$$

for the threshold-gain coefficient required to sustain laser oscillations in a given system. This condition can also be written as

$$g_{th} = g_i + g_m \qquad (2.17)$$

where g_m is the transmission-loss coefficient at the output mirror and is given by eqn (2.16), as

$$g_m = t/2L \qquad (2.18)$$

The existence of this threshold-gain coefficient implies that there is a minimum value of population inversion which must be established before laser action can proceed. This minimum value is called the *threshold* population inversion, N_{th}.

The relation between g_{th} and N_{th} is derived in Appendix B.

The Optical Cavity as an Oscillator

We can see from the foregoing discussion that we can liken the behaviour of a laser to that of an electronic amplifier with positive feedback applied. By analogy, the gain of any amplifier with feedback G_f can be related to its open-loop gain G_0 by the relation

$$G_f = G_0/(1 - \beta G_0)$$

where β is the feedback factor (Fig. 2.18). For oscillation to take place in such an amplifier, the following condition is necessary,

See Horrocks, D.M. *Feedback Circuits and Op Amps* (Van Nostrand Reinhold, 1983).

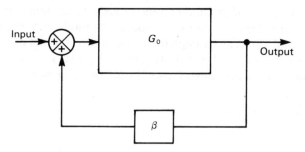

Fig. 2.18 Diagrammatic representation of amplifier with open-loop gain G_0 and positive feedback β.

$$\beta G_0 = 1 \tag{2.19}$$

In other words we have applied *positive* feedback to the system.

From eqn (2.14) the gain of the laser beam as it makes one pass of the medium can be given as $G_0 = \Phi/\Phi_0 = \exp[(g - g_i)L]$. This can be interpreted as the open-loop gain, or gain without feedback, of the system. Comparing eqns (2.14) and (2.18) we see that the two conditions are identical provided that we interpret the feedback factor as,

$$\beta = R^{1/2}$$

and the open-loop gain as

$$G_0 = \{\exp[2(g - g_i)L]\}^{1/2}$$

Substituting into eqn (2.19) and rearranging, we arrive back at the threshold condition as previously given.

Output Coupling

We have discussed above the conditions necessary to sustain laser oscillations at their threshold value in an optical cavity. To obtain a useful output power from the laser by stimulated emission, however, requires that we pump the system with energy at a rate higher than that needed to merely establish the threshold population inversion. The minimum pump power needed to establish a population inversion is the *threshold pump power*, Φ_{th}. The ratio of supplied pump power Φ_{pump} to the threshold pump power is given as

$$p = \Phi_{pump}/\Phi_{th} \tag{2.20}$$

The total radiant power which can be obtained from the laser may be expressed in terms of the power generated in stimulated transitions Φ_{stim} by

$$\Phi_{out} = \Phi_{stim} \left[\frac{g_m}{g_m + g_i} \right] \tag{2.21}$$

where the term in brackets represents the fraction of generated power which emerges from the output mirror. The power generated as stimulated radiation is, itself, related to the pump power supplied to the system less any radiant power emitted in spontaneous transitions, by,

$$\Phi_{stim} = \Phi_{pump} - \Phi_{spont} \qquad (2.22)$$

The radiant power emitted in spontaneous transitions is proportional to the total-loss coefficient $g_m + g_i$ and thus, is related to the threshold pump power by

$$\Phi_{spont} = \Phi_{th}(g_m + g_i)/g_i \qquad (2.23)$$

Hence, substituting eqns (2.20) and (2.22) into eqn (2.22) yields the following relation:

Try this yourself.

$$\Phi_{out} = \Phi_{th} \left[p \left(\frac{g_m}{g_m + g_i} \right) - \frac{g_m}{g_i} \right] \qquad (2.24)$$

Typically, the pump ratio should be at least four times threshold to result in a useful radiant output.

The above relation tells us the output power possible for a given pumping rate and specific loss coefficients. It should be apparent, however, that the power which can be extracted depends on the transmission of the output mirror. Increasing its transmission to extract more power, increases the overall loss of the system, consequently requiring greater pumping power to reach threshold. Reducing the mirror transmission to zero, reduces the threshold pump power to a minimum but prevents any extraction of output. There must, therefore, be an optimum mirror transmission factor at which the radiant output power will be a maximum.

The optimum mirror transmission is obtained by differentiating eqn (2.24) with respect to g_m and equating to zero. We thus arrive at the condition for optimum mirror transmission as

$$g_m(\text{opt}) = g_i(p^{1/2} - 1)$$

Now substituting for g_m from eqn (2.18) we obtain

$$t_{opt} = 2g_i L(p^{1/2} - 1) \qquad (2.25)$$

The maximum flux is obtained by substituting t_{opt} back into eqn (2.24). Hence,

Try solving both these equations for yourself.

$$\Phi_{opt} = \Phi_{th}(p^{1/2} - 1)^2 \qquad (2.26)$$

A neodymium-doped crystal of yttrium-aluminium-garnate (YAG) is to be used as the basis of a laser. The crystal is 30 mm long with an internal-loss coefficient of 1 m^{-1}. Calculate the optimum reflectivity of the output mirror, assuming a pumping ratio of 20.

Exercise 2.4

[0.79]

Summary

In this chapter we have laid the essential foundations required for the rest of our study of optoelectronics. We have shown light to be a complex phenomenon requiring two linked, but distinct, mathematical models to describe its behaviour: the wave model being most use in describing wave propagation, whereas the photon model is needed to describe the emission of light from radiant sources. The emission from such sources is the result of a complex interplay between three processes, namely, absorption, spontaneous emission and stimulated emission of photons. In normal thermal systems stimulated emission is not a dominant process but is found to be crucial to the establishment of population inversions and the subsequent generation of laser light.

In an atomic system containing a specific set of energy levels and enclosed in an optical cavity we can promote stimulated emission at the expense of spontaneous emission. Irradiating such a system with light of the correct frequency raises atoms through intermediate states to the long-life metastable states which can support a population inversion. Spontaneous emission of photons of the desired frequency stimulates the emission of further photons. Those photons travelling parallel to the optical axis of the cavity promote further emission until the gain condition is satisfied. At this point an intense beam of laser light flashes through the partially transmitting mirror.

The laser photons are characterized by bearing a constant phase relationship with each other, by being all of one wavelength and travelling in the same direction in a narrow beam.

Problems

2.1 Calculate the energy of a single photon of 514 nm wavelength. Quote your answer in both SI units and electronvolts.

2.2 If a photon has an energy of 3 eV, what wavelength and frequency does this represent?

2.3 If an electron is accelerated through a potential difference of 100 V, what energy will it acquire?

2.4 Under favourable laboratory conditions, the human eye can just detect about 10^{-18} J of monochromatic light at its most sensitive wavelength of 550 nm, how many photons does this represent?

2.5 How much energy is needed to excite a hydrogen atom into its second excited state? If excitation is carried out by optical means, what wavelength should the irradiating light possess?

2.6 Under good atmospheric conditions, the irradiance which the earth receives from the sun is about 1400 W m^{-2} normal to the sun. Assuming that sunlight is composed entirely of 550 nm photons, what is the rate at which photons arrive at the earth's surface? What pressure do the photons exert on the earth's surface? If the mean orbital radius of the earth is 150×10^9 m, what is the radiant flux emitted by the sun? (Hint: the pressure of light is defined as its irradiance divided by the velocity of light.)

2.7 In the previous example, the average diameter of the sun is 700×10^6 m. Assuming that the sun radiates like a black-body (see Appendix A), what is its surface temperature and its peak emission wavelength?

2.8 If hydrogen gas is in thermal equilibrium with its surroundings at a temperature of 3000 K, what is the ratio of atomic populations between the first and ground states?

2.9 Estimate the mode spacing in a laser cavity of 250 mm length. For a HeNe laser, how many modes are, theoretically, possible?

2.10 The central lasing frequency of a HeNe laser is 474 THz and its spectral linewidth is 1 GHz. Express both these values as wavelengths. Also express the linewidth in units of cm^{-1}.

2.11 Calculate the small signal gain coefficient at threshold for a laser with an internal loss coefficient of 1 m^{-1}, a cavity length of 500 mm and an output mirror transmission of 10%.

2.12 Calculate the optimum reflectivity of the output mirror in the previous laser if pump power is supplied at a rate ten times that needed to reach threshold.

3 Laser Systems

Objectives

☐ To outline the properties of lasers and laser light
☐ To discuss the operation and performance of typical laser systems
☐ To give an insight into aspects of laser design

In the previous chapter we outlined the physical processes necessary to produce amplification of light in an atomic system. Our attention now turns to how we achieve this in practice. What are the constraints imposed in trying to produce a real *optical amplifier*? In other words, how do we make a laser?

Optical Amplifiers: The Laser

The basic requirements in any laser are identical. Firstly, we need a medium which possesses the desired energy level structure to support laser action. The medium may be for instance a gas, a solid insulating crystal, a liquid or a semiconducting crystal. These lines of demarcation also provide us with a means of classifying lasers according to the physical nature of their active medium: hence we have *solid-state* lasers, *gas* lasers, *dye* lasers and *semiconductor* lasers to name some of the best-known classes. Secondly, to establish a population inversion we must pump energy into the system. The method of pumping employed also depends on the active medium. In gas lasers excitation is usually supplied by striking an electrical discharge through the gas; solid-state lasers are often pumped by flooding the crystal with intense radiation from a flashtube; dye lasers can be flashlamp pumped or pumped from another laser; semiconductor lasers are usually pumped via an injection of current to the device. Finally, we need a mechanism by which we can introduce optical feedback and so maintain the gain of the system above all losses. As we saw earlier this is achieved by bounding the medium between two reflectors to form an optical resonant cavity. The complete laser system is shown in Fig. 3.1.

Although many hundreds of different lasers are now available, only a few types are in regular use in engineering. Tables 3.1a–e list the characteristics of the

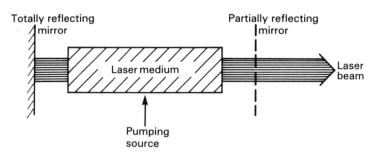

Fig. 3.1 Schematic of laser elements.

Table 3.1a
Principal laser types: gas

Medium	Principal wavelengths	Output	Mode	Typical efficiency %	Typical beam diameter (mm)	Typical divergence (mrad)
HeNe	633 nm	0.1–50 mW	cw	0.1	1	1
Argon	488, 514 nm	5 mW–20 W	cw	0.1	1	1
Krypton	647 nm	5 mW–6 W	cw	0.05	1	1
HeCd	325 nm	1–10 mW	cw	0.1	1	1
	or 442 nm	2–50 mW	cw	10		
CO_2	10.6 μm	20 W–15 kW	cw	10	25	2
CO_2(TEA)	10.6 μm	30 mJ–150 J	50–100 ns pulses up to 1 kHz	5	50	5

Table 3.1b
Principal laser types: solid state

Medium	Principal wavelengths	Output	Mode	Typical efficiency %	Typical beam diameter (mm)	Typical divergence (mrad)
Ruby	694 nm	30 mJ–100 J	10 ns–10 ms up to 3 Hz prf	0.5	5–10	5
Nd-YAG	1.064 μm	10 mJ–150 J	10 ns–1 ms up to 50 kHz prf	1–2	1–10	5
Nd-YAG (Diode pumped)	1.064 μm	1–10 mW	cw	5	1	1
Nd-Glass	1.06 μm	100 mJ–100 J	50 μs–1 ms up to 2 Hz prf	2	3–20	5

principal laser types. We will try to give a broad insight into the design, operation and properties of such lasers. We should bear in mind that modern laser design is more complex than outlined here. We are only attempting to get a feel for what is involved. For example, in all our analyses so far we have assumed an optical cavity bounded by parallel reflectors. In practice many different mirror arrangements are adopted.

For further information see the *Laser Handbook* Vol. 1 (North-Holland, 1984).

Table 3.1c
Principal laser types: semiconductor

Medium	Principal wave-lengths	Output	Mode	Typical efficiency	Typical divergence
GaAlAs	750–905 nm	1–40 mW	cw	up to 20%	10° × 30°
In GaAsP	1.1–1.6 nm	1–10 mW	cw	up to 20%	10° × 30°
Phase-coupled arrays	790–850 nm	100 mW–1 W 1 W–10 W	cw	20–40%	10° × 10°

Table 3.1d
Principal laser types: excimer

Medium	Principal wave-lengths (nm)	Average output	Mode	Typical effi-ciency	Typical beam diameter (mm)	Typical diver-gence (mrad)
Argon Fluoride	193	50 W	5–25 ns pulses up to 1 kHz prf	up to 1%	2 × 4 to 25 × 30	2–6 mrad
Krypton Fluoride	248	100 W	2–50 ns pulses up to 500 Hz	up to 2%	2 × 4 to 25 × 30	
Xenon Fluoride	351	30 W	1–30 ns pulses up to 500 Hz	up to 2%	2 × 4 to 25 × 30	

Table 3.1e
Principal laser types: dye

Medium	Principal wave-lengths	Output	Mode	Typical effi-ciency %	Typical beam diameter (mm)	Typical diver-gence (mrad)
Flash lamp pumped	340–940 nm tuneable	up to 50 W ave	200 ns–4 μs pulses, up to 50 Hz	up to 1	5–20	0.5–5
Ion laser pumped	400–1000 nm tuneable	up to 2 W	cw	5–25	0.6–1	1–2
Pulsed laser pumped	300–1000 nm tuneable	up to 15 W ave	3–50 ns pulses up to 10 kHz	depends on pump light	2–10	0.36

Properties of Lasers

To enable us to compare the common laser systems, it would be instructive here to outline the principal properties and parameters of lasers and laser light.

Radiant Power (Flux)

We should by now appreciate that lasers are amplifiers of light. One of their great virtues is that they can deliver high radiant power whether this be in pulsed or continuous mode. Pulsed power ranges from a few watts produced by semiconductor lasers to around 10^{18} watts delivered by solid-state lasers used in laser fusion systems. Continuous optical power ranges from a few milliwatts in HeNe lasers to several kilowatts delivered by CO_2 lasers.

Because, as we will see later, laser beams tend to deliver their high output flux in a very narrow directional beam, it is sometimes appropriate to discuss the output in terms of the *irradiance*, or flux per unit area incident on a surface.

Optical power is often called *flux*, see Chapter 5.

Irradiance will be discussed in Chapter 5.

Worked Example 3.1

A typical laboratory HeNe laser emits 5 mW of light in a beam of about 1 mm diameter. Calculate the irradiance received at a nearby surface if the beam is shone on it.

Solution: The irradiance at the surface is given by,

E = incident power/irradiated surface area

$\quad = \Phi/A$

$\quad \sim 5 \times 10^{-3}\,\text{W}/(1 \times 10^{-3}\,\text{m})^2$

$\quad \sim 5000\,\text{W m}^{-2}$

To put this figure into perspective the irradiance received at the earth from the sun is around $1400\,\text{W m}^{-2}$. Argon lasers with their inherently higher power can deliver irradiances of about a thousand times that of the sun!

Coherence

One of the most striking features of gas lasers is the emission of a narrow pencil of scintillating light which is used to dramatic effect in laser-light shows. The high degree of parallelism, directionality and monochromaticity is seen to some extent in all lasers and is related to the *coherence* of the beam. Two types of coherence may be distinguished namely, *temporal* which defines the phase constancy of the beam between two given instants in time and *spatial* which defines the phase constancy of the beam between two points across the wavefront (Fig. 3.2). The above parameters are in turn influenced by the design of the optical cavity and the energy structure of the laser medium.

Coherence Length

For practical purposes temporal coherence is best described in terms of the *coherence length* of the laser. If the output of a laser beam is split into two parts and recombined after travelling two different paths of the same length, the two beams will *interfere* to form an interference pattern at the point of recombination.

The phenomenon of optical interference was outlined in Chapter 1 and is discussed by Hecht, E. and Zajac, A. *Optics* (Addison-Wesley, 1974).

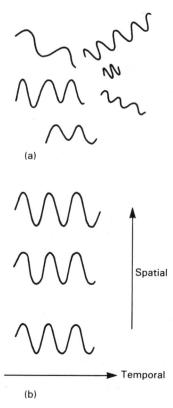

(a)

Spatial

Temporal

(b)

Fig. 3.2 Coherent and incoherent light. (a) Spatially and temporally incoherent. (b) Spatially and temporally coherent.

Changing the path length of one beam with respect to the other will reduce the visibility of the interference pattern until at some point, when the path lengths differ by $\pm l_c$, no interference occurs. The parameter l_c defines the coherence length of the laser.

Coherence length is itself related to the monochromaticity or linewidth of the beam by,

$$l_c = c/\Delta f \qquad (3.1)$$

The narrower the linewidth the greater is its degree of coherence.

Linewidth is also, as we saw earlier, related to the number of longitudinal modes which can be supported within an optical cavity. By constraining the laser to operate in a single longitudinal mode we can reduce its linewidth and hence increase its coherence length. Linewidths for gas lasers operating multimode are typically tens of gigaHertz leading to coherence lengths of a few millimetres. By operating in a single longitudinal mode linewidths come down to a few-hundred megaHertz giving rise to coherence lengths of some metres. By contrast the greater linewidths of solid-state lasers give coherence lengths of fractions of a millimetre in multimode, and a metre or so in single mode.

Refer to 'single-frequency operation' in Chapter 2.

Exercise 3.1

Calculate the coherence length of an argon laser operating multimode with an overall linewidth of 10 GHz.

[30 mm]

It is in applications such as holography that coherence length is of great importance since it effectively determines the distance over which two mutually coherent waves can travel and still produce an interference pattern.

Holography will be discussed in detail in Chapter 6.

Beam Divergence

The spatial coherence of a laser beam is defined by the number of transverse cavity modes which the cavity can support and is in turn related to the angular spread, or *divergence*, of the beam as it exits the cavity. Angular spread is due to the diffraction, or bending, experienced by light as it passes through an aperture. The beam divergence $\Delta\theta$ or the amount by which the beam diameter increases over a given distance is least for the lowest-order modes. A laser oscillating in the fundamental TEM_{00} mode will therefore exhibit greatest spatial coherence and least divergence. Divergence is usually measured in radians. For example a divergence of 1 mR corresponds to a 1 mm increase in diameter over 1 m of travel. Divergence is least for gas lasers at around 0.5 mR increasing through solid-state lasers at about 5 mR to the 20 mR or so for semiconductor lasers.

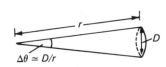

Beam divergence

Also related to divergence is the ability to focus a laser beam to a small spot size. The diameter d to which a laser beam can be focused by a lens of focal length f is often expressed as

$$d = 2f\Delta\theta \tag{3.2}$$

Smallest spot sizes are therefore obtained for lasers operating in the fundamental TEM_{00} mode.

Returning to the optical disk system in Chapter 1, we can see that for the high-bit densities needed to encode video signals the original video disk players used HeNe lasers because of the ease with which they can be focused to small spots. By contrast the CD system has a lower requirement on bit densities and can utilize the semiconductor laser with its cost and size advantage.

Gas Lasers

Gas lasers are characterized by having as their active medium an atomic, ionic or molecular gas with the desired set of energy levels. The energy level structure of a low-density gas approximates to that of the isolated atom with transitions occurring between individual levels or perhaps some closer-spaced groups of levels. The medium is enclosed in a cylindrical tube and sealed at each end by a mirror to form the optical cavity. Pumping is via an electrical discharge passed through the gas.

The physics of gas discharges is complex but fundamental to the operation of gas lasers. Useful texts include Hemenway, C.L. Henry, R.W. and Caulton, M. *Physical Electronics* (Wiley, 1968), *The Encyclopaedia of Physics* (Van Nostrand Reinhold USA, 1985) ed. Besancon, R.M.

The HeNe Laser

Historically, the HeNe laser was the second working system to be demonstrated but was the *first* working *gas* laser and was also the first to produce a *continuous* output beam. It has since become the workhorse of the laser world and finds extensive use in, for example, general optics, holography, surveying and image processing.

The active medium is a gaseous mixture of helium and neon, in roughly 10:1 proportion, contained in a closed quartz tube. A glow discharge is created in the gas by applying a high voltage of between 1 and 10 kV between a pair of electrodes

The HeNe laser was developed by Ali Javan and his team at Bell Laboratories, USA in 1960, 3 months after the ruby laser. Read the original paper,

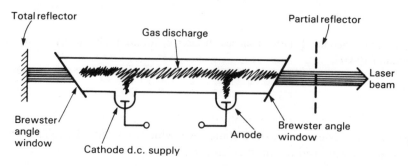

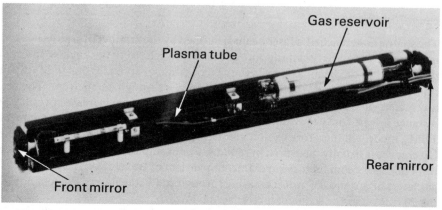

Fig. 3.3   HeNe laser (photograph shows construction of Spectra Physics 25 mW laser type 107A).

Javan, A. Population Inversion and Continuous Optical Maser Oscillation in a Gas Discharge containing a He-Ne Mixture, *Physics Review Letters*, Vol. 6, p. 106, 1961.

Words like 'lasing' and its corresponding verb 'to lase' are becoming accepted additions to the scientist's vocabulary!

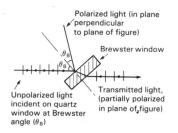

inserted at opposite ends of the tube (Fig. 3.3). Once struck, a steady d.c. current of typically 3 to 10 mA is sufficient to keep the discharge established. The electric current which flows leads to excitation of the He atoms due to collisions with the energized electrons. The excited He atoms in turn transfer some of this energy through atomic collisions to the Ne atoms which are thereby raised to their upper excitation levels (see the energy level diagram in Fig. 3.4). A population inversion is established in this level and laser action at a wavelength of 632.8 nm can take place. It is important to note that the He atoms provide the means to excite the Ne atoms; lasing action takes place in the Ne levels.

Laser action will continue as long as population inversion can be maintained and a continuous beam is emitted with powers in the range 0.5 to 50 mW. The gain of the HeNe laser at this wavelength is low and only small cavity losses can be tolerated, implying the use of high-quality mirrors with low scattering and absorption losses. Linewidths in gas lasers are much lower than for any other type of laser system. For the HeNe laser in particular, linewidths are typically of the order of a few picometres.

In cheap, low-power HeNe lasers of up to around 1 mW, the mirrors are fixed directly to the discharge tube. In the high-power versions, the mirrors are external to the discharge tube and the tube must be sealed by fitting windows. These windows are inclined at a specific angle to the optical axis, the *Brewster angle*. When unpolarized light is incident on a piece of glass or quartz inclined at its Brewster angle only the components of light polarized in the plane of the window

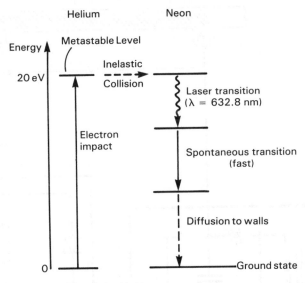

Fig. 3.4   HeNe energy diagram.

will be transmitted. Other components will be reflected and the light emerges *plane polarized*.

Pump Power Requirements for a HeNe Laser

Low-power gas lasers like the HeNe are usually pumped from a high-voltage d.c. supply across the ends of a gas-filled tube. Typically a circuit such as that shown in Fig. 3.5 will be sufficient to give long life and optimum performance provided some care is taken in selection of operating parameters. The most important consideration is the negative voltage-current relationship which is typical of these lasers (Fig. 3.6). The negative slope of this curve dictates the use of a limiting ballast resistor to provide a positive resistance and prevent overloading the power supply. Initially, a high voltage of around 8 kV (V_{start}) is needed to produce sufficient ionization of the gas such that lasing will start. Lasing cannot be sustained until a particular current I_{lase} is reached where the positive resistance of the ballast resistor balances the negative resistance of the gas column. The current through the tube is increased from this point until the optimum operating current is reached where the output power is a maximum. The optimum operating voltage V_{op} for the tube shown, is around 1800 V at an optimum current I_{op} of 7.5 mA. Suitable values for the load resistor, of about 50 to 100 kΩ, are recommended by the manufacturer as well as optimum operating points for voltage and current.

Worked Example 3.2

A typical HeNe laser tube has the negatively sloped current-voltage relation shown in Fig. 3.6. Calculate the value of a suitable ballast resistor to drive this laser from the circuit shown in Fig. 3.5.

Solution: The onset of lasing action occurs at about the mid range of the characteristic curve yielding I_{op} = 4.5 mA and V_{op} = 1.6 kV. Drawing a tangent to the curve through this point enables the ballast resistor to be calculated as follows,

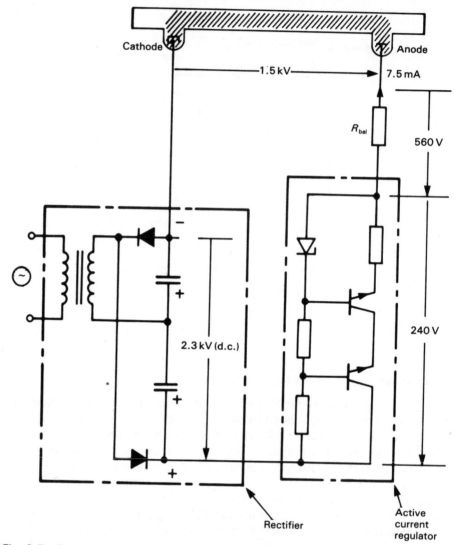

Fig. 3.5 Typical power supply for HeNe laser (adapted from Goldsborough, J.P.
Design of Gas Lasers, in *Laser Handbook* Vol. 1 (North-Holland)).

$$R_{bal} = \Delta V / \Delta I$$
$$= (1700 - 1400)/V(7 \times 10^{-3} - 3 \times 10^{-3})A$$
$$= 75 \text{ k}\Omega$$

For a more detailed discussion of gas-laser power supplies see Weil, T.J. in 'Electronics and Power' or Goldsborough, J.P. in 'Design of Gas Lasers' *The Laser Handbook* (North-Holland, 1984) ed. Arrechi, A.T. and Schulz-DuBois, Vol. 1.

Argon lasers, as we will see next, are inherently more complex because of the need to supply large amounts of pump power to initially ionize the argon atoms and then excite the ions to the required energy levels. The design of such lasers is outside the scope of this text.

The Argon-ion Laser

Unlike the HeNe laser, in which lasing takes place in the atomic species, the active

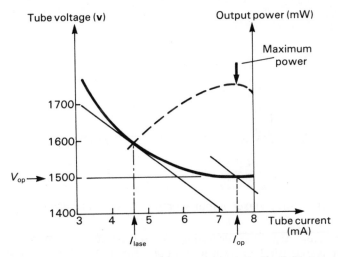

Fig. 3.6   Typical voltage-current characteristic for HeNe laser.

medium in the argon laser is a plasma of excited *ions*. An electric discharge is created in a narrow tube of gaseous argon. Argon atoms are first ionized and then excited by multiple collisions with electrons into their upper energy levels (Fig. 3.7). It requires about 16 eV to ionize the argon atoms and *another* 20 eV to excite them into their higher laser levels. Because of the existence of a band of closely-spaced upper levels, several laser transitions occur simultaneously in the blue-green region of the spectrum, the strongest being at 514 and 488 nm (Fig. 3.8). Due to the high energy required to ionize and excite the argon atoms, very high current densities are needed, of order 1 A mm⁻².

Returning briefly to atomic theory, an ion is an atom which has one or more of its orbital electrons excited entirely out of the influence of the nucleus. The ion has a net positive charge. Ions consisting of an excess of electrons will have a net negative charge.

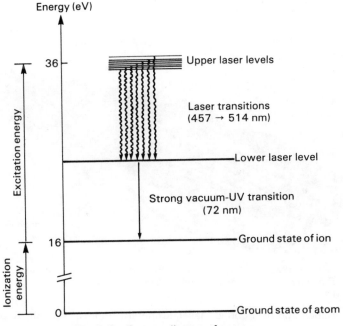

Fig. 3.7   Energy diagram for argon.

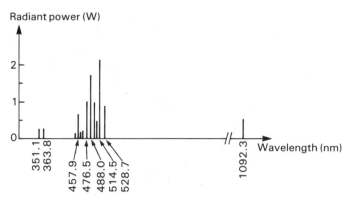

Fig. 3.8   Argon laser output spectrum (9 W total radiant power)

Exercise 3.2   A typical laboratory Argon laser delivering 5 W cw (continuous wave) radiant flux would need a three-phase electrical supply of 30 A per phase at 415 V and associated water cooling to operate it. What is the conversion efficiency of the system?

[n 0.02%]

A magnetic field surrounds the laser tube to help constrict the gas discharge and keep the current density high. The longitudinal field increases the electron density in the plasma by constraining the electrons to move in a helical path around the field lines. This prevents loss of electrons to the walls. The discharge tube is normally made of a material with a low thermal conductivity such as beryllium oxide (BeO), graphite or a metal-ceramic tube construction. To keep running-temperatures low, the latest generation of argon lasers feature metal discs inserted inside the tube to act as heat exchangers. The layout of a typical commercial argon laser is shown in Fig. 3.9.

A new technology has arisen out of the manufacture of laser tubes.

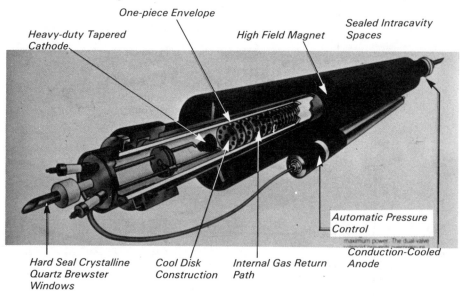

Fig. 3.9   Schematic of argon-ion laser (Coherent 'Innova' series). (Reproduced by permission of Coherent (UK) Ltd.)

Argon lasers emit around 1 to 20 W of flux distributed amongst all the lasing wavelengths. As much as 5 or 6 W can be obtained at the most powerful of these wavelengths: the 514 nm line. Argon lasers are commonly used at a single wavelength rather than in *all-lines* mode. Single-wavelength operation is achieved by inserting a *wavelength-selective* component, such as a prism, into the cavity. Rotating the prism brings each wavelength into operation as appropriate. Having selected single-wavelength operation the user may then choose to go to *single-frequency* operation which is essential for applications demanding a long coherence length. This is accomplished by the insertion of a *frequency-selective* component, such as an etalon. The sequential process of moving from all-lines to single-frequency operation, with its consequent output power penalty, is shown in Fig. 3.10.

Generally the user would select the optimum operating wavelength for a specific application.

See Chapter 2.

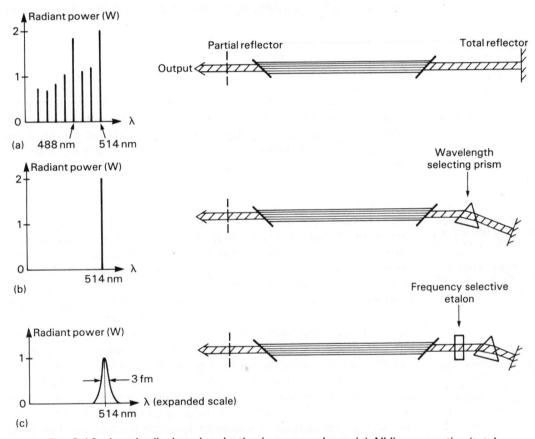

Fig. 3.10 Longitudinal-mode selection in an argon laser. (a) All-lines operation (total power 6 W). (b) Single-line operation (2 W). (c) Single-frequency operation (1 W).

Common uses of argon lasers are in holography, eye surgery, spectrochemistry, optical image processing, semiconductor processing and last, but not least in terms of numbers of lasers supplied, laser-light shows.

The Carbon Dioxide Laser

The third gas laser we will discuss is fundamentally different from the other two. The important energy levels are provided not by the distribution of electrons but are due to the wiggling and jiggling of the entire carbon dioxide molecule itself. The CO_2 molecule can be pictured as a linear arrangement of O-C-O atoms which vibrate in relation to each other. Several different modes of vibration give rise to a set of energy levels with transitions far into the infra-red. The principal CO_2 wavelength is 10.6 μm. Continuous power outputs up to several kilowatts are obtainable, making this laser the favoured choice for materials-processing applications such as cutting, welding and annealing. Unlike most other gas lasers, the CO_2 has an appreciably high efficiency, typically 10 to 25%. To reach the high powers required from these lasers, cavity lengths can stretch to 2 or 3 metres or more. Variants, such as the Transversely Excited Atmospheric (TEA) in which excitation is applied across rather than along the discharge, enable more compact CO_2 lasers to be built. Other variations such as waveguide CO_2 lasers can produce hundreds of milliwatts in a laser no more than two- or three-hundred millimetres long.

For more details see Vol. 1 of the *Laser Handbook*.

Solid-State Lasers

To help avoid confusion in terminology with the *semiconductor* laser, solid-state lasers are sometimes now referred to as *doped-insulator* lasers.

Solid-state lasers are characterized by having as their active medium, not a gas, but a solid rod or slab of crystalline insulator doped with a small amount of impurity. It is the impurity constituent which provides the required energy structure to produce laser action. The crytalline lattice primarily acts as a host material but also influences the overall energy structure.

This behaviour need not concern us here.

Ruby Laser

Theodore Maiman, of Hughes Labs, showed the first working laser to the world in 1960. See Maiman, T. Stimulated Optical Emission in Ruby Lasers, *Nature* Vol. 187, p. 493, 1960 for the original account.

The ruby laser takes its place in history by being the first working laser to be demonstrated. The active medium is a cylindrical crystal of synthetic ruby (Al_2O_3) doped with roughly 0.05% by weight of chromium ions (Cr^{3+}). The ends of the rod are polished flat and parallel. High standards of flatness and parallelism are demanded: the flatness over the entire end face should vary by no more than a quarter of a wavelength ($\frac{1}{4}\lambda$) and both surfaces should be parallel to within a few seconds of arc. Pumping is provided by optical energy from a flashtube (Fig. 3.11). Often the ruby and flashtube are placed at the respective foci of an elliptical reflector to ensure that as much light as possible is pumped into the rod. The ruby absorbs pumping energy in the blue-green region of the spectrum and emits its principal laser energy at a wavelength of 694.3 nm. The absorption of green light raises the chromium ions to a broad band of upper levels from which it relaxes very quickly to the upper level of the laser transition (Fig. 3.12). This is a metastable state with a 3 ms lifetime. A population inversion is established between this state and the ground state.

Other pumping geometries have been used including Maiman's original idea of using a helical flashtube wrapped around the rod. More common now is the close-coupled geometry.

Radiant emission occurs as a short pulse of energy over a duration of around 250 μs. Typical energies for ruby lasers range from a few millijoules to several-hundred joules. The output pulse actually occurs in a series of relaxation spikes superimposed on the fluorescent background (Fig. 3.13).

The Nd-YAG Laser

Now superceding the ruby as the most widely used doped-insulator laser, the host material is a crystal of yttrium-aluminium-garnate (YAG) doped with neodymium

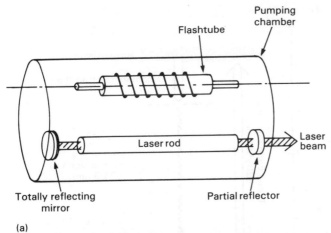

(a)

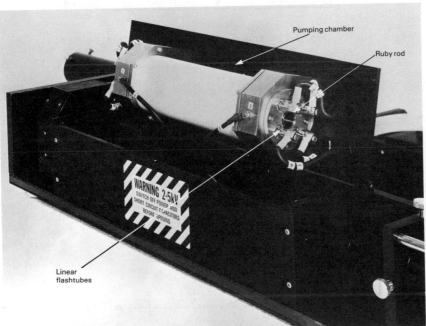

Fig. 3.11 Solid-state laser. (a) Schematic. (b) Four-lamp, close-coupled pumping chamber.

(Nd^{3+}) ions. The advantage of Nd-YAG over ruby is its lower threshold and higher gain. The implications of this are that the laser runs cooler, is more efficient in operation, can be made smaller for a given energy and can be operated at relatively high repetition rates up to 50 kHz.

Optical pumping raises the ions into the large number of levels around 2 eV, with non-radiative relaxation into the upper laser level (Fig. 3.14). Laser emission takes place at $1.064\ \mu m$.

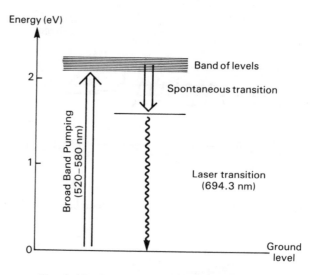

Fig. 3.12   Energy diagram for ruby (Cr^{3+}).

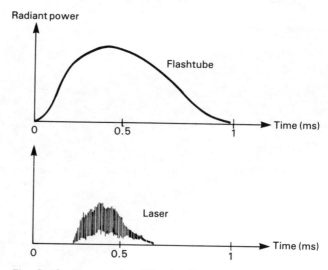

Fig. 3.13   Temporal profiles for flashtube and ruby laser.

Q-switching

Sometimes also known as 'burst', 'relaxation-oscillation' or 'normal' mode.

The mode of operation we have just described for solid-state lasers is known as *free-running*. There is an alternative mode of operation, commonly used, known as *Q-switched* (QS) which enables much higher peak powers to be obtained for a given pump energy.

In the QS mode an optical shutter is placed in the cavity, usually between the back mirror and the laser medium, such that feedback from the mirror is prevented. In this way the ability of the cavity to store energy is increased over its normal value and, even although a population inversion is still building up, laser action is prevented. The ability of a resonant cavity to store energy is usually defined in terms of its Q factor. The Q of a cavity is a measure of how much energy

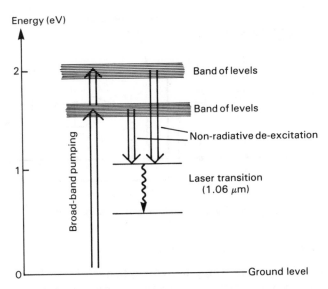

Fig. 3.14   Energy diagram for Nd-YAG (Nd^{3+}).

can be stored in it against the power loss from it. A cavity with a high Q enables a lot of energy to be stored. Lasers typically have Q's of a few million compared to a few hundred for electrical oscillators. The population inversion will build up far above its normal threshold value. When the shutter is opened, emission of energy proceeds in one short, sharp burst. Pulse durations as short as a few nanoseconds are obtainable in this way giving rise to average powers of megawatts and higher.

The alternative name for this mode of operation is *giant pulse.*

Several mechanisms exist for Q-switching. One of the simplest, and earliest used, methods was the replacement of the total reflector, the back mirror, with a rotating mirror. Laser oscillation is prevented until the mirror comes into alignment with the optical axis of the cavity, in other words, until front and back mirrors are momentarily parallel. Better control of the operation with, hence, shorter pulses and higher flux is obtained with the use of a bleachable organic dye, such as *cryptocyanine*. The dye is initially opaque and withholds laser oscillations until a certain flux is reached, whereupon, the dye becomes transparent and laser action can proceed. The particular value of flux threshold depends on type and thickness of dye and the wavelength of laser light. The main disadvantage of the dye is its limited lifetime and the variability in the timing of the onset of laser action. These shortcomings are overcome by the use of a *Pockel cell*. When a high voltage is placed across a Pockel cell the plane of polarization of the light beam is rotated. Cavity oscillations cannot take place between beams of different polarizations thus laser action is suspended. On removal of the voltage the plane of polarization reverts to its original state and oscillation proceeds.

Such dyes tend to be carcinogenic: another disadvantage!

Calculate the average flux of a Na-YAG laser giving an energy of 100 mJ in the free-running mode over a pulse duration of 200 μs (FWHH). If a Pockel's cell inserted into the cavity reduces the pulse duration to 10 ns (FWHH) for the same energy output, what is the new average flux which can be obtained? Calculate also the number of photons emitted in each case.

[500 W, 10 MW]

Exercise 3.3

FWHH means that the pulse duration referred to is the *Full Width* measured at *Half* its peak *Height.*

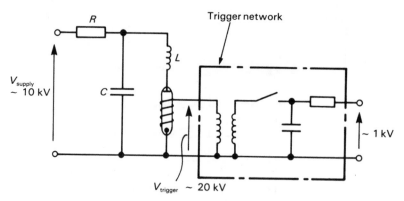

Fig. 3.15   Typical flashlamp pumping circuit.

Flashlamp Pumping of Solid-State Lasers

The principal method of supplying pump energy to a solid-state laser is by flooding with broad-band optical radiation from a linear flashtube. Energizing is accomplished by dumping the stored energy from a charged capacitor across the ends of the flashtube. The type of circuit used depends on the nature of the application. If a single laser pulse is required then a simple RC network supplied from a high voltage would normally suffice. For fast repetition rates of up to 10 Hz then a critically damped LC network gives better control of the charging and discharging conditions. The energy dumped into the flashtube is given by

$$W_{flash} = \tfrac{1}{2} CV^2$$

where C is the capacitance and V is the charging voltage. Triggering is from a high-voltage pulse applied to a coil of wire wrapped around the quartz envelope. A typical circuit is shown in Fig. 3.15.

Worked Example 3.3   Making reasonable assumptions calculate the energy required from a flashtube to produce a population inversion in a Nd-YAG laser.

Solution: We show in Appendix B that at an overall loss of 1 m^{-1}, a population inversion of around 3.1×10^{21} m^{-3} is needed between the laser levels to achieve threshold in a Nd-YAG crystal. In a 4-level system, like Nd-YAG, the energy required to create the inversion is equivalent to that needed in raising atoms to the fourth level of the system, since we assume that decay from this level to the meta-stable level is rapid. From Fig. 3.14, the pump energy per atom is about 1.5 eV and the total required energy density is

$$
\begin{aligned}
dW_{pump}/dV &= (N_u - N_l)\, W_{pump} \\
&= 3.1 \times 10^{21}\ \text{m}^{-3} \times 1.5\ \text{eV} \times 160 \times 10^{-21}\ \text{J eV}^{-1} \\
&= 744\ \text{J m}^{-3}
\end{aligned}
$$

Taking a typical crystal of 30 mm length by 3 mm diameter the total input energy needed is

$$
\begin{aligned}
W_{in} &= 744\ \text{J m}^{-3} \times 30 \times 10^{-3}\ \text{m} \times \pi(3 \times 10^{-3}\ \text{m})^2/4 \\
&= 158\ \mu\text{J}
\end{aligned}
$$

The light emitted from the flashtube covers a broad-band spectrum from about 400 nm to 1.4 μm. Only a small percentage of this light, say about 1%, will coincide with the useful pumping bands and only about 5% of this will actually be absorbed by the crystal. Additionally, we have to take into account the conversion efficiency of the lamp; typically about 30% of the electrical input to the flashtube will be converted into optical energy. Hence, we arrive at a value of

$$W_{flash} = 158\ \mu J/(0.01 \times 0.05 \times 0.3)$$
$$= 1.05\ J$$

for the energy which needs to be dumped into the flashtube to reach threshold.

Since at threshold the system will only just be lasing, the pumping rate should be increased over its threshold value in order to extract some useable power. Although there are some relations which will enable us to calculate the pumping energy needed for optimum output, a simple rule of thumb tells us that a pumping energy of 20 times threshold should suffice. Hence, we need the flashtube to deliver about 20 J into the system.

(See Chapter 2).

Having calculated how much energy we need to discharge through the flashtube we now need to estimate the circuit parameters which will enable us to achieve this. Since the voltage-current curve of a flashlamp is non-linear, empirical design procedures are used to calculate suitable circuit parameters. We will outline this process by way of the following example.

Design a charging network to deliver 20 J to the Nd-YAG system of Worked Example 3.3.

Worked Example 3.4

Solution: Taking the critically damped LC network shown in Fig. 3.15 as the basis of our design, we have first of all to choose a suitable flashtube duration. We know that the spontaneous lifetime of the upper state of the laser transition is 200 μs. We have already discussed the need to ensure that the rate of population of the upper laser level is faster than the rate at which it depopulates. Hence, the flashtube duration should be significantly less than the spontaneous lifetime of the laser transition. Taking then the flash duration t_{flash} to be about a quarter of the spontaneous emission lifetime we arrive at a suitable flash duration of 50 μs.

(See Chapter 2).

The total pulse duration for a critically damped circuit is given by

$$t_{flash} = \pi(LC)^{1/2} \qquad (3.3)$$

In a critically damped circuit the resistance, inductance and capacitance are related by

$$1/LC = (R/2L)^2 \qquad (3.4)$$

yielding the value for R as

$$R = (4L/C)^{1/2} \qquad (3.5)$$

While the lamp is conducting, the effective resistance of the circuit is the internal resistance of the lamp itself given by

$$R_{int} = \rho l_{arc}/A_{arc} \qquad (3.6)$$

The origins of the formulae used here can be seen in most texts on electricity and magnetism such as Bleaney B. and Bleaney, B.I. *Electricity and Magnetism* (Oxford University Press, 1976) or Smith, R.J. *Circuits, Devices and Systems* (Wiley, 1984).

where ρ is the resistivity of the lamp and is typically of the order 0.2 mΩ m, l_{arc} is the arc length and A_{arc} is its cross-sectional area. In choosing a suitable arc length for the flashtube we must bear in mind the length of the crystal which we have already chosen to be 30 mm. An arc length, therefore, of 35 mm will ensure that as much energy as possible is pumped into the crystal while at the same time keeping energy wastage to a minimum. For a tube bore of 3 mm we get an approximate value for R_{int} of 0.85 Ω. Hence, combining eqns (3.3), (3.4) and (3.5) we arrive at circuit parameters of

$$C = 36 \ \mu F$$

and

$$L = 6.8 \ \mu H$$

Finally, the charging voltage is

$$\begin{aligned} V &= (2 \ W_{flash}/C)^{\frac{1}{2}} \\ &= (2 \times 20 \ J/36 \times 10^{-6} \ F)^{\frac{1}{2}} \\ &\sim 1 \ kV \end{aligned}$$

On this basis we should be able to build such a laser.

Modern Solid-State Lasers

The principles and procedure we have just outlined should enable us to design and build a simple Nd-YAG laser, although commercially available lasers are naturally somewhat more complex. In practice, rather than grow longer and longer rods and designing correspondingly bigger capacitor banks, modern laser design is more subtle. Usually a primary laser rod, known as the oscillator, producing some 10 to 30 mJ is built. The output from the oscillator is timed to arrive at a second rod, the amplifier, just as the flashtube energy has created a population inversion in it. The incoming pulse stimulates the emission of photons giving rise to an amplified output of a joule or so. Additional stages of amplification can be added as necessary. Arrangements such as this also allow better control of the temporal and spatial profile of the beam. A typical layout for a ruby laser is shown in Fig. 3.16.

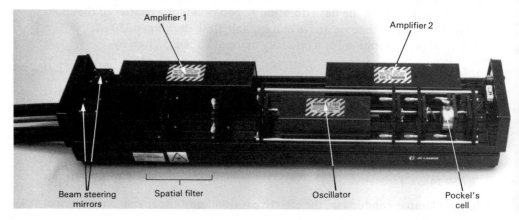

Fig. 3.16 Commercial ruby laser for holographic applications. (Reproduced by permission of Lumonics Ltd.)

Semiconductor Lasers

The semiconductor laser is in many ways the ultimate optoelectronic source. By providing high power in a small package at low cost the laser diode, to give it its alternative name, has become the standard source for optical communications and high-density storage applications like the optical disk. To appreciate its operation fully, we need to look for a little while at some basic semiconductor principles in so far as they apply to laser diodes.

Some Basic Semiconductor Theory

The energy structure in a semiconductor is somewhat different from the crystalline insulators discussed earlier. In a pure semiconductor, the difference in energy between the valence band and the conduction band is much less than that of an insulator, of the order 1 eV as compared to several electronvolts. Whenever an electron is excited out of the valence band across the energy gap and into the conduction band, it leaves an ion with a vacant bond. We usually consider this vacant bond as being a *hole* and carrying an amount of positive charge equal in magnitude to that of the electron. Holes are free to move about the crystal in much the same way as electrons. De-excitation of the electron causes it to lose its excess energy as heat or, more importantly for us, as a photon of light. In so doing the electron returns to the valence band where it recombines with a hole. By adding specific dopants to the pure material, two types of semiconductor can be fabricated: those with an excess number density of electrons (negatively charged *n-type*) and those with an excess number density of holes (positively charged *p-type*) as seen in Fig. 3.17. Single bulk semiconductors, however, cannot produce the conditions necessary for efficient light production. In practice we need to fabricate a semiconductor with a differential doping structure: a *pn-junction*.

Although in reality a pn-junction diode is fabricated by selectively doping a single slice of semiconducting material so that it is part p-type and part n-type, it helps us to describe what happens if we imagine the device as being made from separate slices of p-type and n-type brought together in intimate contact (Fig. 3.18). On contact, those electrons and holes which are close to the junction have sufficient energy to *diffuse* across to the other side where they can combine

For a more complete discussion of semiconductors, see Sparkes, J.J. *Semiconductor Devices* (Van Nostrand Reinhold, 1987).

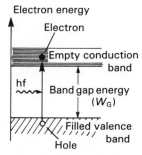

Creation of an electron-hole pair by optical excitation

The number density of electrons is equal to the number of electrons per unit volume. Similar for holes.

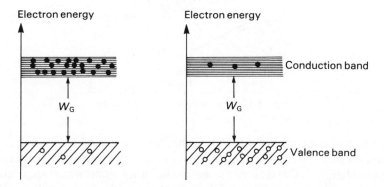

Fig. 3.17 Energy band diagrams for p- and n-type semiconductors. (a) 'n-type' semiconductor showing electron excess in conduction band. (b) 'p-type' semiconductor showing excess of holes in valence band.

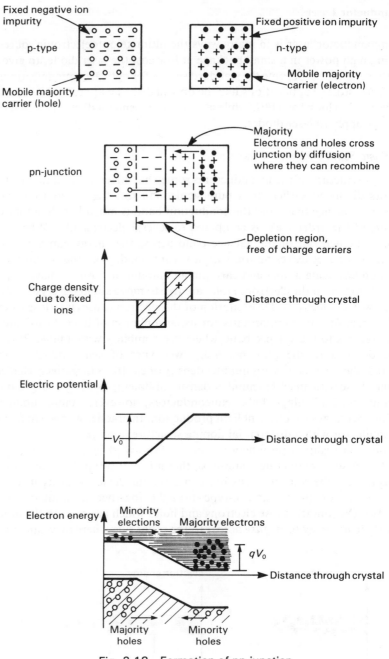

Fixed negative ion
impurity

p-type

Mobile majority
carrier (hole)

Fixed positive ion impurity

n-type

Mobile majority
carrier (electron)

pn-junction

Majority
Electrons and holes cross
junction by diffusion
where they can recombine

Depletion region,
free of charge carriers

Charge density
due to fixed
ions

Distance through crystal

Electric potential

$-V_0$

Distance through crystal

Electron energy

Minority
electrons

Majority electrons

qV_0

Distance through crystal

Majority
holes

Minority
holes

Fig. 3.18 Formation of pn-junction.

Diffusion is the term used to
describe the movement of
particles by virtue of a
concentration gradient.

Remember, the point of highest
electric potential corresponds to
the most positively charged

with carriers of the opposite polarity. A thin layer is created on either side of the
junction, the *depletion layer*, which is empty of free charge carriers. Because there
are no free charges in this region, the ionized impurity atoms which are fixed in the
lattice ensure that there is a potential gradient across the depletion layer directed
from the n-side to the p-side. The presence of this charged layer acts to keep
electrons in the n-side and holes in the p-side. In other words, any further move-

ment of majority carriers across the junction is inhibited. This holds for all majority carriers whether they are thermally generated or introduced via doping. On the other hand electron-hole pairs are continually being thermally generated on either side of the junction. Electrons which are created on the p-side, that is minority carriers, will be pulled across the junction to the n-side by virtue of the potential gradient. The same will be true for the minority holes on the n-side. This leads to an equilibrium situation where the number of electrons crossing from n to p by diffusion is balanced by those crossing from p to n by drift, with the end result being no net flow of current across the junction. The magnitude of the equilibrium potential across the junction is V_0. A diagram representing the energy level structure of the composite pn-junction can be drawn as in Fig. 3.18 where qV_0 is the magnitude of the barrier expressed in terms of energy.

region. From electromagnetic field theory, electric field $\mathscr{E} = dV/dx$.

When electrons are in the n-region they are known as *majority* carriers, if they cross to the p-type they become *minority* carriers, and vice versa for holes.

Movement of charge carriers by virtue of an applied electric field is called *drift*.

Injection Luminescence

Let us now consider what happens when a current is injected through the diode (Fig. 3.19). On applying a forward voltage across the ends of the diode the equilibrium situation is disturbed. The energy barrier is reduced by an amount qV_{app} equivalent to the applied potential energy, $q(V_0 - V_{app})$. Majority carriers of both kinds now find it easier to cross the junction. Because the barrier energy has decreased, the diffusion current must now exceed the drift current giving rise to a net flow of current from the p-side to the n-side. This current is known as the *injection current*.

The p-type is made positive with respect to the n-type.

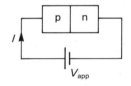

On crossing the junction, carriers can recombine with those of the opposite polarity and spontaneously emit radiation. The longest wavelength which can be emitted, corresponding to an electron dropping from the bottom of the conduction band to the top of the valence band, is given by

$$\lambda_c = hc/W_G$$

where W_G is the energy of the band gap. Shorter wavelengths are emitted when the electron drops from higher energy states in the conduction band.

In a normal pn-junction diode spontaneous radiation represents lost energy and the diode is designed to minimize this loss. In our situation we want to exaggerate this effect. The phenomenon of injection luminescence forms the basis of both the light-emitting diode (LED), and the laser diode, although to produce laser light we still need to create a population inversion.

We will outline the operation of LEDs later in this chapter.

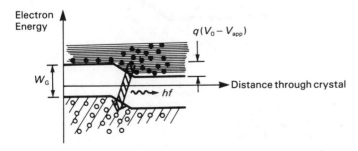

Fig. 3.19 Injection luminescence in a semiconductor diode.

The Laser Diode

It was constructed in 1962 at General Electric, USA. See Hall, R.N. *et al.* Coherent Light Emission and GaAs Junctions. *Applied Physics Letters* Vol. 1, p. 62, 1962.

The first operational laser diode consisted of a single crystal of gallium arsenide (GaAs), doped to form a pn-junction and a forward potential applied (Fig. 3.20). Gallium arsenide was chosen rather than silicon because of its *direct* band gap. In direct band-gap materials, conduction electrons can lose energy directly by photon emission. In *indirect* band-gap materials, the electrons have first to lose excess momentum before emitting a photon. Direct band-gap materials are therefore more efficient at producing light.

To create a population inversion and enhance the possibility of recombination, high levels of doping are needed to ensure that, in the depletion region, filled states in the conduction band are directly above empty states in the valence band. This applies only across a very narrow region of the depletion area, about 1 nm wide, known as the *active layer*. For significant gain, a high current density of order several-hundred amps per square millimetre is necessary. The onset of lasing is characterized by a specific injection current known as the threshold current. Below this threshold, I_{th}, light emission will be spontaneous and incoherent. The ends of the diode are cleaved and polished to provide the optical cavity. The sides are roughened to prevent loss of light. These early lasers had lifetimes of only a few hundred hours and required cooling by liquid nitrogen for efficient operation.

Improvements to the performance of laser diodes came with modifications to the simple single-junction (*homojunction*) structure. By surrounding the active layer with regions of lower refractive index than the active layer itself, laser emission was horizontally confined to this narrow junction region where it could stimulate more photons (Fig. 3.21). This is a kind of waveguide effect. Such *single heterojunction* (SH) structures can produce short laser pulses with peak powers as high as a few watts for injection currents in the 1 to 40 amp range. Further improvements came with the *double heterojunction* (DH) diode. Reducing the active region even further and sandwiching between a double layer brought threshold currents down to hundreds of milliamps and permitted efficient operation in both pulsed and con-

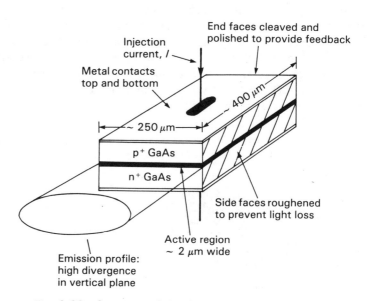

Fig. 3.20   Structure of simple homojunction laser diode.

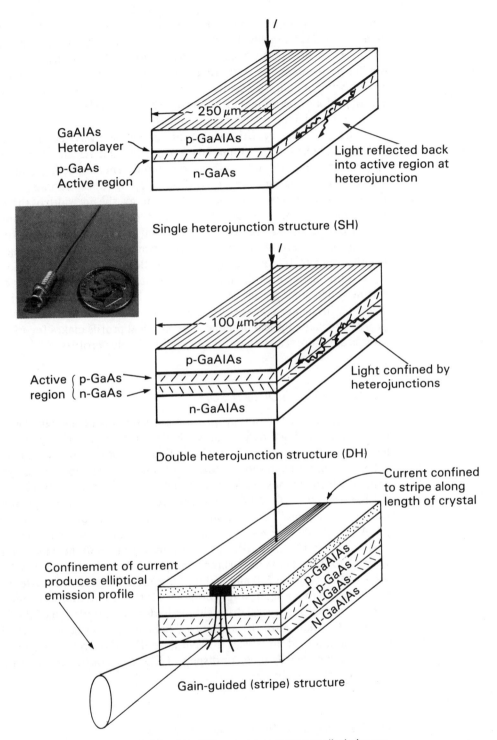

Fig. 3.21   SH, DH and stripe-structure diode lasers.

Further details on these structures can be found in Hecht, J., *High Technology* (1984) pp. 43–50.

tinuous modes. Improvements to the quality of the output beam came with the *index-guided* laser in which the beam is confined both vertically and horizontally by layers of different refractive index. *Gain-guided* (or *stripe*) lasers achieve a similar performance by restricting the current flow through the diode, and hence the emission of light, to a central stripe. The lower injection currents, tens of milli-amps, prolong operational life. Contrast this with the few hundred amps needed to operate the original homojunction diodes.

Diodes are available with a wide range of emission wavelength in the near infra-red, typically 820 nm, 850 nm, 904 nm, 1.3 μm and 1.5 μm. The different wave-lengths are obtained by using a compound semiconducting material such as $Ga_xAl_{1-x}As$ and varying the ratio of Ga:Al doping concentrations. Because optical transitions occur between broad bands of energy, the linewidth of the emitted light is larger than for other types of laser and is typically around 5 to 10 nm. The small length of the optical cavity, of order 500 μm, pushes the longitudinal-mode spacing up to several-hundred gigaHertz and, consequently, brings coherence lengths down to fractions of a millimetre. Because the laser radiation is emitted from a tiny active area diffraction of the beam gives rise to higher divergences. Furthermore, since the active area is several times wider than it is high, divergence is greater in the horizontal plane than in the vertical plane giving rise to a spatial profile which is elliptical in shape. The elliptical profile makes focusing of a diode laser beam more tricky than for those with circular profiles.

The Light-Emitting Diode (LED)

We can now deviate slightly from our discussion of coherent light sources to bring in the light-emitting diode. By dispensing with the optical cavity and reducing the level of doping compared with the laser diode, the radiation emitted from the diode will result mainly from spontaneous transitions. As such, low-intensity radiation is emitted which will possess neither spatial nor temporal coherence.

Like laser diodes the LED is based on a chip of semiconducting material such as GaAs. Compound semiconductors such as gallium arsenide phosphide $(GaAs_xP_{1-x})$ allow selection of the band-gap width and hence, emission wave-length, by varying the As:P ratio. For pure GaP ($x = 0$) the band gap is 2.26 eV, whereas for pure GaAs ($x = 1$) the band gap is about 1.44 eV providing a range of wavelengths between 550 and 860 nm. Construction is slightly different from the laser diode in that a shallow junction is made to allow as much radiant emission as possible to escape. Several different methods of encapsulation of the junction are employed to maximize the amount of light which can be emitted. The type of encapsulation used influences the spatial profile of the output beam.

Although primarily used for displays LEDs can sometimes prove a useful low-cost alternative to the laser diode.

Driving Circuits for Diode Lasers and LEDs

In designing driving circuits for diode lasers, account must be taken of the low resistance of the diode when operated with a forward voltage across it. The implication here is that the diodes should be fed from a current source, i.e. one having a high internal resistance. Such conditions can be realized by replacing the

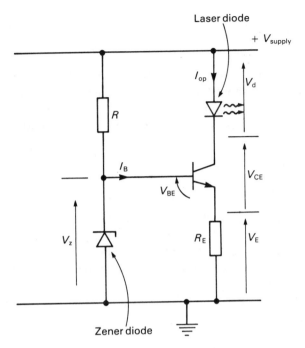

Fig. 3.22 Driving circuit for laser diode.

load resistor in a fully stabilized common emitter circuit by the laser diode (Fig. 3.22). The constant current is obtained by substituting a zener diode in place of the resistor, normally between base and ground. The current through the laser diode is given by

$$I_D = (V_z - V_{BE})/R_E$$

where V_z is the breakdown voltage of the zener diode, V_{BE} the base-emitter voltage and R_E the emitter resistance. An alternative arrangement is to place the diode in the emitter leg of the circuit. Because the emitter current is determined by the base voltage, and hence the zener voltage, this arrangement helps to insure against supply voltage variations.

The design of such circuits has been explained in detail by Ritchie, G.J. *Transistor Circuit Techniques* (Van Nostrand Reinhold, 1983).

Worked Example 3.5

Design a circuit to drive a type C86000E diode so that it delivers around 7 mW of radiant flux.

Solution: From Table 3.2, the threshold current for this diode is 75 mA with a forward voltage drop of 2 V. Using the circuit shown in Fig. 3.22 as the basis of our design and the design rules according to Ritchie we can set the emitter voltage as

$$V_E \sim \tfrac{1}{5} V_{supply}$$

For a supply voltage of, say, 15 V we arrive at a value of 3 V for V_E. To achieve the typical output flux of 7 mW we need to operate the diode somewhere above its threshold current of 75 mA. Table 3.2 gives a value of 200 mA for the maximum forward current, so an operating current of around 100 mA should suffice. Hence, the emitter resistance is given by

$$R_E = V_E/I_{op}$$
$$= 3 \text{ V}/100 \text{ mA}$$
$$= 30 \text{ } \Omega \tag{3.7}$$

Taking a lower standard resistor value of 27 Ω should ensure that the diode is operated well above threshold. Substituting this value back into eqn (3.7) we obtain an actual operating current of 111 mA.

Now to calculate the required value of zener voltage, we need the base voltage which is given by

$$V_B = V_z = V_{BE} + V_E$$
$$= 0.7 \text{ V} + 3.0 \text{ V}$$
$$= 3.7 \text{ V}$$

Since the voltage drop across the laser diode is 2 V, this leaves about 10 V still dropped directly across the collector-emitter of the transistor. We must be careful that we do not exceed V_{CEmax} for the transistor. The final bias resistor R_1 is chosen according to the gain h_{FE} of the transistor. Assuming h_{FE} to be around 150, we obtain a base current of

$$I_B = I_{op}/h_{FE}$$
$$= 111 \text{ mA}/150$$
$$= 740 \text{ } \mu\text{A}$$

Thus the base resistor is given by

$$R_B = (V_{supply} - V_z)/I_B$$
$$= (15 - 3.7) \text{ V}/740 \text{ } \mu\text{A}$$
$$\sim 15 \text{ k}\Omega$$

The circuits shown here are very simple. In practice greater control of operating parameters is necessary to ensure reliable operation. For further information and some circuit-design, data sheets such as those from RCA, TRW, Toshiba or ITT are useful starting points. See also Texas Instruments, *Optoelectronics* (McGraw-Hill 1978).

The fully stabilized common-emitter circuit also provides a useful starting point if the diode is to be driven in pulsed mode with the modulating signal capacitively coupled to the base. The operating point is determined by the emitter resistor and the potentiometer bias chain, R_1 and R_2 (Fig. 3.23) and should be chosen around the mid-point of the output power-current characteristic. If this information is not available select the operating point at midway between the threshold current and the peak forward current.

Summary

A laser consists of three main constituents: an active medium in which we can hope to produce a population inversion between a specific pair of energy levels or energy bands; an optical cavity so that we can introduce an element of feedback which will balance system gains against system losses; and finally a means of pumping energy into the system. The precise configuration of the laser depends on the nature of the active medium and the source of pumping energy. Lasers are classified into groups according to their active medium.

The most common industrial lasers are: gas lasers, typified by HeNe, argon-ion and CO_2; solid-state lasers as characterized by ruby and Nd-YAG; and semi-

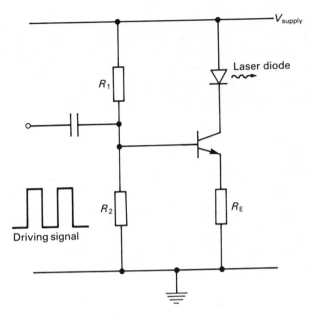

Fig. 3.23   Modulating circuit for laser diode.

conductor lasers. The design of suitable pumping sources for such lasers is generally complex but simple approaches can be applied in certain cases to provide the basis of some provisional designs.

Problems

3.1 In a HeNe laser, Ne atoms are excited into the upper level of their laser transition at 20.66 eV above the ground. Emission of a laser photon takes the atom to an energy of 18.70 eV, before further de-excitation by spontaneous photon emission to the 16.84 eV level. What are the photon wavelengths emitted in the laser and spontaneous transitions?

3.2 An argon-ion laser emits 2.4 W of cw radiation at its strongest emission line of 514 nm. What is the rate of photon emission?

3.3 A hypothetical laser has an energy structure composed of 3 levels, an upper state 2.5 eV above ground, a long-lived state 1.8 eV above ground and the ground state itself. If the laser is optically pumped, what wavelength of light should the pumping source have and what is the wavelength of the laser transition?

3.4 A pulsed Nd-YAG laser emits 150 mJ of energy in a single pulse of 15 ns (FWHH) duration. What is the average pulse power and how many photons does the pulse contain, assuming that all light is emitted in the 1.064 μm transition?

3.5 Assuming that a population inversion of 116×10^{21} m^{-3} is required to sustain a ruby rod at its threshold value, calculate the total energy required to establish this condition. Assume the ruby rod to be 50 mm long by 5 mm diameter. (If necessary, make use of the energy level diagram for ruby and the data given in Worked Example B.1.)

3.6 Making sensible assumptions, estimate the output energy required of a flashtube to bring the laser in the previous example to its threshold.

3.7 Following the procedure outlined in Chapter 3, design a charging network to pump the laser in Example 3.6 to four times above its threshold. State clearly any assumptions and design rules you have employed.

3.8 Estimate the threshold population density for a HeNe laser with the following parameters:

$$g_{th} = 1 \text{ m}^{-1}$$
$$t_{spont} = 100 \text{ ns}$$
$$\Delta f = 1 \text{ GHz}$$
$$n = 2.4$$

3.9 Design a circuit to drive an LED with the following characteristics:

forward current = 50 mA
forward voltage drop (at specified current) = 1.9 V
radiant flux = 25 μW (cw)

Table 3.2
Laser diode type RCA C86000 E

Material	GaAlAs	Emission wavelength	820 nm
Junction	DH	Line width	4 nm
Threshold current	75 mA	Max. output power	10 mW(cW)
Typical forward current	100 mA	Response time	< 1 ns
Max. forward current	200 mA	Forward voltage drop	2 V

Photodetectors

☐ To classify photodetectors according to the mechanism by which they respond to light

☐ To define the parameters by which the performance of photodetectors may be compared and selected

☐ To describe the operation, performance and application of some common detectors

---

In the design or operation of an optoelectronic system, the question soon arises as to how we 'see' and monitor the light passing through. What is needed is some means of electronically detecting and measuring the radiant energy being transmitted from the source. This brings us to a consideration of the final link in our chain: the photodetector.

Classification of Photodetectors

As will soon become apparent, there is a wide range of photodetectors available to the engineer each with differing properties and performance. To help us compare and assess the performance of individual detectors, we can classify them according to the mechanism by which they respond to incident light. The photodetectors most commonly encountered in optoelectronics fall into three main categories, namely:

(i) semiconductor detectors, like the junction diode or light-dependent resistor, in which electrons are excited from the valence band to the conduction band of the photosensitive material;

(ii) photoemissive detectors, as characterized by the photomultiplier tube, in which electrons are ejected from a photosensitive material on irradiation by light;

(iii) thermal detectors, such as the thermopile, which rely on the heating effect of light to raise the temperature of the irradiated material, with the subsequent change in one or more of its electric properties.

This is an example of the well-known photoelectric effect, first satisfactorily explained in 1905 by Einstein, for which he was awarded the *Nobel Prize*.

Sometimes known as the external photoelectric effect.

The selection of the most suitable photodetector obviously depends on its particular performance and the application for which it is intended. For most purposes in optoelectronics the semiconductor detector reigns supreme. Its low cost, small size, ruggedness and modest power requirements, coupled with its broad spectral range, good sensitivity and fast response, make it almost the ideal optoelectronic detector. There are occasions, however, when the semiconductor detector is not the best choice. Photomultipliers although bulky, fragile and requiring high-voltage power supplies, are indispensible when the light source is weak. When we are down to a few microwatts of optical power, the PM tube becomes attractive. Additionally, speed of response for PM tubes can be in the picosecond range. For measurement of optical radiation out to the far infrared, we have to consider the use of thermal detectors, particularly if the incident flux is of the order of several watts and above. Speed of response and sensitivity is, however, poor.

Performance Parameters of Photodetectors

To help us choose the most suitable detector for any particular application, we need to define some parameters which will allow us to compare their performance on equal terms. Over the next few pages we will outline some of the more useful pieces of photodetector 'language' encountered in the data sheets.

Spectral Response

The first characteristic of interest to us is the spectral response of the detector. Since we are measuring optical radiation, stretching from the UV to the far IR, few parameters have the same magnitude over more than a narrow region of this spectrum. All the parameters which follow must be specified with respect to specific wavelengths. The consequence of this behaviour is that it is essential to try to match as closely as possible the peak-emission wavelength of the source with the peak response of the detector. This is particularly important if optimum performance or detection of low light levels is required. For example, it is of no use trying to measure the output flux of a CO_2 laser with a detector whose peak sensitivity is in the visible part of the spectrum.

Quantum Efficiency (η)

The quantum efficiency of a photon detector is a measure of how many photo-electrons are produced for every photon incident on the photosensitive surface. Hence, quantum efficiency is defined as

$$\eta = (n_e/n_p) \times 100\% \tag{4.1}$$

where n_e is the rate of photoelectron generation and n_p is the incident photon rate. The ideal situation of 100% quantum efficiency, i.e. one photoelectron for one photon, does not usually occur. In practice, values in the range 5 to 30% are more typical.

Responsivity or Sensitivity (S)

The magnitude of the electrical signal output from a photodetector in response to a particular light flux is often given in terms of its *responsivity* or *sensitivity S* which is defined as

$$S = i_d/\Phi$$

or

$$S = V_d/\Phi \tag{4.2}$$

where i_d is the output current of the detector, V_d is the output voltage and Φ is the incident light flux.

The photon-generated current can be derived by considering what happens when a beam of light of wavelength λ irradiates a photodetector. The rate of incident photons arriving at the photodetector is given by dividing the total light flux by the energy of a single photon, that is

$$n_p = \Phi / W_{ph}$$
$$= \Phi \lambda / hc$$

The rate of photoelectron production is obtained by multiplying the incident photon rate by the quantum efficiency η of the photosensitive surface. Hence,

$$n_e = \eta \Phi \lambda / hc$$

Consequently, by combining the photoelectron production rate with the charge on one electron, we obtain the generated photocurrent as,

$$i_d = (\eta \Phi \lambda / hc)q \tag{4.3}$$

From eqn (4.2) we arrive at the responsivity of the photodetector as,

$$S = \eta \lambda q / hc \tag{4.4}$$

Calculate the responsivity of a photosensitive material with a quantum efficiency of 1% at 500 nm.
$$[4.0\,mA\ W^{-1}]$$

Exercise 4.1

Noise Equivalent Power

All detectors, regardless of type, produce a small but measurable output signal even in complete darkness. The presence of this background signal, or noise, sets a lower limit to the intensity of light which can be detected. To be seen by the detector the incident light needs to produce an output greater than that of the noise signal. In photomultipliers and semiconductor devices, the background signal is thermally generated: a few electrons are excited into the conduction energy levels to produce a background current known as the *dark current*. The lower the operating temperature of the device the lower is its dark current. Dark currents are typically in the picoamp to nanoamp region.

To help specify the minimum detectable power of a detector we use a quantity known as the *noise equivalent power* (NEP). The NEP is defined as the radiant flux which produces an output signal equal in magnitude to that produced by the noise signal. Thus, unlike dark current, NEP gives a measure of the minimum amount of light which can just be detected.

Specification of NEP is not straightforward but varies with detector bandwidth, area and temperature. To remove the dependence on bandwidth, NEP is usually quoted as an inverse function of the square root of the bandwidth. Good detectors have NEP values of around 10^{-12} to 10^{-14} W Hz$^{-\frac{1}{2}}$.

The units of NEP are W Hz$^{-\frac{1}{2}}$.

This is not yet the end of the story. To allow detectors to be compared regardless of their sensitive area, a normalized, reciprocal NEP is often quoted and called the *specific detectivity (D*)*. Generally the higher the value of $D*$ the better is the freedom from background noise.

The units of $D*$ are m Hz$^{\frac{1}{2}}$ W^{-1}.

A word of warning, both NEP and $D*$ appear in data sheets apparently at random. Be sure you know which one is being quoted. Generally, you want *low* NEP but *high* $D*$.

Response Time

Finally, we come to the response time of the detector. All detectors have an inertia in their reaction to incident light. The response time is a measure of how long it

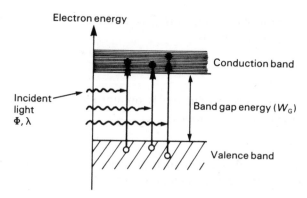

Fig. 4.1   Detection of light by LDR.

takes a detector to respond to a change in light flux falling on it. Response times are usually measured with reference to a square input pulse and both rise and fall times are often quoted. As a working rule you would normally choose a detector with a rise time of about a tenth of the shortest pulse duration to be detected. For example, to measure the temporal profile of a 50 ns laser pulse, we would need to choose a detector with a rise time of 5 ns or less.

Light-Dependent Resistor

Alternative names are the *photoresistor* or *photoconductor*. The latter names, however, can cause confusion with the pn-junction diode operated in *photoconductive* mode. Watch out for inconsistency in the literature.

We have briefly discussed the absorption of light by a semiconductor in Chapter 3.

The light-dependent resistor (LDR) is the first semiconductor detector we will consider. The LDR consists of a slab of bulk semiconducting material such as cadmium sulphide (CdS) or cadmium selenide (CdSe) with a pair of electrical contacts across its ends. As light falls on the semiconductor, electron-hole pairs are produced within it. Provided that the energy of the incident photon is greater than the band gap of the material, the photoelectron will gain enough energy to be excited into the conduction band (Fig. 4.1). The increased number of electrons in the conduction band, with a corresponding increase of holes in the valence band, produces an increase in the conductivity and hence a decrease in the resistivity of the material. Usually an electric potential in series with a load resistor is applied across the semiconductor to pull the electrons and holes to their respective terminals (Fig. 4.2). The band gap of the material sets an upper cut-off wavelength above which no electrons will be ejected into the conduction band. This wavelength is given by

$$\lambda_c = hc/W_G \tag{4.5}$$

where W_G is the band-gap energy of the semiconductor.

Worked Example 4.1
Calculate the cut-off wavelengths of intrinsic CdS, CdSe and PbS LDRs, if their respective band-gap energies are 2.4, 1.7 and 0.4 eV.

Solution: For CdS, the band-gap energy in Joules is,

$$W_G = 2.4 \, eV \times 160 \times 10^{-21} \, J \, eV^{-1}$$
$$= 384 \times 10^{-21} \, J$$

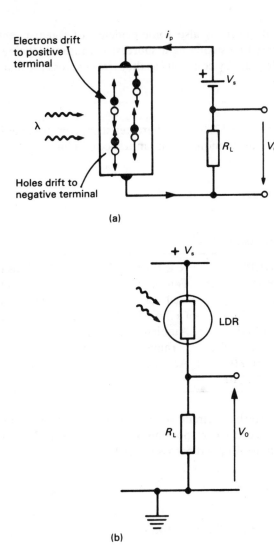

(a)

Electrons drift
to positive
terminal

i_p

$+ V_s$

R_L V_o

λ

Holes drift to
negative terminal

$+ V_s$

LDR

R_L V_o

(b)

Fig. 4.2 Operation of LDR. (a) Schematic of operation. (b) Circuit diagram.

Therefore, the cut-off wavelength (from eqn (4.5)) is

$$\lambda_c = hc/W_G$$
$$= (663 \times 10^{-36}\ \text{Js} \times 300 \times 10^6\ \text{m s}^{-1})/384 \times 10^{-21}\ \text{J}$$
$$= 518\ \text{nm}$$

This detector will respond only to light in the blue-green part of the spectrum.

Similarly to CdSe, λ_c is 731 nm and will respond to all visible wavelengths, whereas PbS with a cut-off at 3 μm is sensitive out to the infra-red.

To increase the cut-off wavelength, the above materials are sometimes doped with foreign donor atoms which reduce the effective band gap.

Response times of LDRs depend purely on the drift of the photon-generated carriers to their respective electrodes and are, therefore, relatively poor with 50 ms

The influence of doping on the effective band gap of a semiconductor is discussed by Seymour J. *Electronic Devices* (Pitman) and by Anderson J.C. *et al. Materials Science* (Van Nostrand Reinhold, 1985).

being fairly typical. There is also some evidence of signal retention for a few seconds if illuminated by strong light sources. Because the charge carriers create more electron-hole pairs by collisions as they travel across the material, gains can be as high as 10^5.

Worked Example 4.2 A typical LDR (type ORP 12) has the characteristics given in Table 4.1. Using a 9 V battery, design a circuit to monitor the emittance of a light beam in the range 1 to 1000 lx.

Solution:

Photometric units such as lux are discussed in Chapter 5.

Table 4.1
Light Dependent Resistor

Type	ORP 12	Rise time	75 ms (50 lx to dark)
Manufacturer	Radiospares	Fall time	350 ms (dark to 50 lx)
Cell resistance		Cell voltage	110 V (dc max)
at 20 000 lx	10 Ω	Cell dissipation	220 mW
1000 lx	150 Ω	Ambient temperature	−10 to 60°C
50 lx	2.4 kΩ	range	
1 lx	80 kΩ	Active area	0.6 cm²
dark	≥ 10 MΩ		

Let us now adopt the basic circuit shown in Fig. 4.2 and set the minimum voltage we would like to record (corresponding to 1 lx) at 100 mV. Thus at minimum irradiance, the voltage dropped across the cell is

$$V_c = V_s - V_L$$
$$= 9\text{ V} - 0.1\text{ V}$$
$$= 8.9\text{ V}$$

The current through the cell is therefore

$$I_c = V_c/R_c$$
$$= 8.9\text{ V}/80 \times 10^3\ \Omega$$
$$= 111\ \mu\text{A}$$

and the corresponding load resistance is

$$R_L = V_L/I_c$$
$$= 0.1\text{ V}/111 \times 10^{-6}\text{ A}$$
$$= 900\ \Omega$$

Taking the next highest standard resistance value, we have,

$$R_L = 1\text{ k}\Omega \text{ (standard)}$$

At 1000 lx, the cell resistance is 150 Ω, hence the total resistance between the supplies in 1150 Ω. So the cell current is given by

$$\dot{I}_c = V_s/(R_c + R_L)$$
$$= 9\text{ V}/1150\ \Omega$$
$$= 7.8\text{ mA}$$

The corresponding voltage across the load is therefore

$$
\begin{aligned}
V_{L(max)} &= R_L \times I_{c(max)} \\
&= 1000\,\Omega \times 7.8 \times 10^{-3}\,A \\
&= 7.8\,V
\end{aligned}
$$

The Junction Photodiode

The basic general-purpose photodiode is no more than a pn-diode with its junction exposed to incident light (see Chapter 3 for an introduction to the operation of the pn-junction diode). Under equilibrium conditions, i.e. with no applied external potential or illumination, an energy barrier qV_0 exists across the depleted areas on either side of the pn-junction (Fig. 4.3). The barrier effectively prevents diffusion of majority carriers across the junction, except for a small number whose energy is greater than that of the barrier. This small amount of majority-carrier diffusion is balanced by an equal amount of minority-carrier drift in the opposite direction by virtue of the electric field across the junction. The minority carriers are thermally generated and 'fall over' the potential barrier to opposite sides of the junction. The majority-carrier current from p to n is negated by the minority-carrier current from n to p. The result being no net current flow through the diode.

Under illumination, this equilibrium is upset: electron-hole pairs are created in the depletion region and are swept apart by the potential difference across the junction. Electrons are pulled into the conduction band of the n-type, and holes into the valence band of the p-type. At this point two distinct modes of operation are possible depending on whether the diode is operated with no applied voltage, the *photovoltaic* mode; or with an applied reverse voltage, the *photoconductive* mode.

Photovoltaic Mode

In the photovoltaic mode the diode is operated open circuit. The electron-hole pairs which are generated separate and drift to opposite sides of the depletion layer.

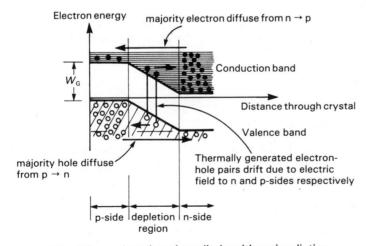

Fig. 4.3   pn-junction photodiode with no irradiation.

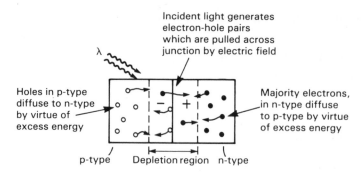

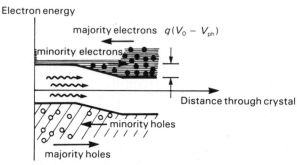

Fig. 4.4   Irradiated pn-junction photodiode in photoconductive mode.

Since the electrons are pulled to the n-side and the holes to the p-side, a photon-induced reverse current i_{ph} flows through the diode from the n-side to the p-side. The effect of this behaviour is that the energy barrier is reduced with respect to its equilibrium value (Fig. 4.4). Now, more majority carriers are able to cross the junction, that is, holes cross from p to n and electrons cross from n to p, creating a forward current through the diode. Since the diode is open circuit, the photon current must exactly balance the forward current. No net current flows and, therefore, the drop in the energy barrier is seen as a forward voltage across the ends of the diode. It is this signal which is measured and gives this mode of operation its name, *photovoltaic*.

The net current which flows through the diode is therefore given as the difference between the forward current i_f due to the induced forward bias and the reverse photocurrent i_{ph}. The total, we have already seen is zero, hence,

$$i_T = i_f - i_{ph} = 0 \qquad (4.6)$$

The forward current i_f is expressed by the *diode equation* as,

$$i_f = i_0[\exp(qV_{ph}/kT) - 1] \qquad (4.7)$$

The diode equation appears in simplified form in Ritchie, G.J. *Transistor Circuit Techniques* (Van Nostrand Reinhold, 1983) p. 5 and its derivation appears in Sparks, J.J. *Semiconductor Devices* (Van Nostrand Reinhold, 1987).

The diode equation is the fundamental relation between the forward current produced in a pn-junction for a particular applied potential. Hence, combining eqns (4.6) and (4.7), we obtain

$$i_r = i_0[\exp(qV_{ph}/kT) - 1] - i_{ph} = 0 \qquad (4.8)$$

Assuming the exponential term to be much greater than unity, we have

$$i_0 \exp(qV_{ph}/kT) \sim i_{ph} \qquad (4.9)$$

The externally measurable photovoltage V_p across the ends of the diode is therefore given by

$$V_{ph} = (kT/q)\ln(i_{ph}/i_0) \qquad (4.10)$$

We have shown earlier, in eqn (4.4), that the photon-generated current in a detector is a linear function of the light flux,

$$i_{ph} = \eta\Phi\lambda q/hc$$

and hence, the voltage developed across an open circuit diode is a logarithmic function of the flux, as shown below,

$$V_{ph} = (kT/q)\ln(\eta\Phi\lambda q/i_0 hc) \qquad (4.11)$$

Therein lies the main disadvantage of the photovoltaic mode in light detection: the relation between the incident light and the output signal is non-linear. Furthermore since charge carriers are only detected as they drift, influenced by the internal field, to their respective contacts, speed of response depends on diode thickness and is generally slow. The principal benefit to be gained from photovoltaic operation is low noise due to the absence of a leakage current.

Photoconductive Mode

When a pn-junction is operated under reverse potential, that is if the positive battery terminal is connected to the n-side and the negative terminal to the p-side, electrons in the n-side are pulled out of the depletion region and holes are pulled from the p-side. This leaves more fixed ions of both kinds in the depletion region causing it to widen. Consequently, the energy barrier increases in accordance with the applied potential (Fig. 4.5). The flow of majority carriers of any kind is halted and the only current which can flow is the reverse current i_0 due to thermally-generated minority carriers.

Under illumination, the photogenerated electron-hole pairs are again swept apart by the internal electric field across the junction and constitute a reverse photon current i_{ph} in the same direction as the thermally-generated leakage current.

We see now the major benefit of the photoconductive mode: it is the photon-generated current which constitutes the measured output signal and not the voltage drop across the diode. Hence, we have an output signal which is a linear function of incident light flux where

$$i_{ph} = \eta\Phi\lambda q/hc \qquad (4.12)$$

Photoconductive operation results in a higher response speed than photovoltaic. Because of the wide depletion layer and, consequently, higher electric field, the transit time for charge carriers to reach their respective electrodes is reduced. The main disadvantage of the PC mode is the increased noise due to the ever-present leakage current.

The current-voltage response of an irradiated pn-junction is shown in Fig. 4.6, for both photovoltaic and photoconductive modes. With no illumination, the response of the diode is as shown on the characteristic curves and corresponds to the situation described by the diode equation (4.7). By increasing the forward voltage, the forward current through the diode will increase as shown, whereas, under reverse potential the only current which flows is the reverse leakage current i_0. On increasing the irradiance, the reverse photon current increases to i_{ph} and the

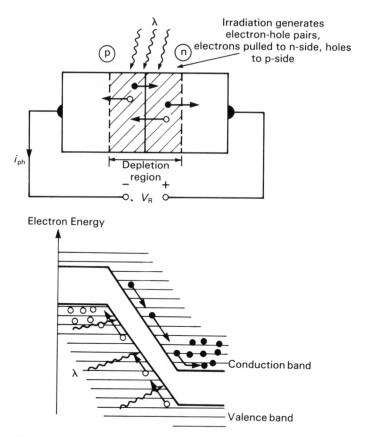

Fig. 4.5   Irradiated pn-photodiode in photoconductive mode.

whole curve shifts downwards by this amount. The forward voltage drop across the open circuit diode for a given irradiance is given by the point at which the curve intersects the voltage axis at $i = 0$. For a given reverse voltage, say V_R, the near linear increase in i_{ph} with irradiation can be seen.

The pin Diode

By modifying the basic pn-junction structure to include a wide layer of intrinsic material between the p and n types, we arrive at what is known as a 'p-i-n' diode. The presence of the intrinsic layer maintains the electric field over a wider than normal depletion layer. Because donor and acceptor impurity concentrations are low, the resistivity is high. Therefore only a small reverse bias is needed to increase the width of the depletion layer until it stretches the full distance between the terminals. This is the *fully depleted* situation. The pin diode has a very fast response time, nanoseconds or less, and is beginning to replace the basic photodiode as the 'standard' photodetector for general-purpose applications.

The Avalanche Photodiode

Avalanche diodes also produce fast response times, but unlike pin diodes, a large amount of internal amplification is obtained. The basic pn-junction is highly

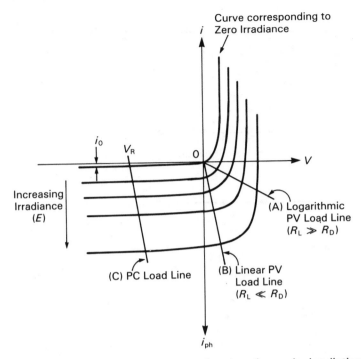

Fig. 4.6   Current-voltage response of pn-junction under irradiation.

doped and operated at a high reverse bias, pushing the diode over the knee of the breakdown limit. Carriers crossing the depletion region gain enough energy to produce further carriers by avalanche multiplication. Avalanche diodes benefit from the inclusion of a 'guard-ring': an isolation ring around the active area which helps to reduce the leakage current when biased to the same voltage as the diode and restrict the avalanche effect to the middle of the illuminated area.

Photodiode Circuits

Before designing a circuit for photodiode operation, we need to know at the outset whether we want photoconductive or photovoltaic operation: in other words, do we need linearity and high speed or low noise and high sensitivity. To assist our design process, the current-voltage characteristics of the diode under illumination such as those shown in Fig. 4.6 are helpful. These curves show the downward shift of the characteristics as the irradiance increases.

The equivalent circuit representing a photodiode is shown in Fig. 4.7. The major components are the photon-generated current i_{ph}, forward current through the diode i_f, a noise current i_0, the series resistance of the diode R_s and a dynamic resistance R_D. The dynamic resistance varies between 500 kΩ and 100 MΩ depending on diode area and construction, and is an important parameter in choosing the correct circuit values.

The simplest circuit for photovoltaic operation is merely a series diode-resistor combination (Fig. 4.8a). The load resistor should be chosen to be greater than the diode dynamic resistance (say 10 GΩ) to ensure that the load line lies almost

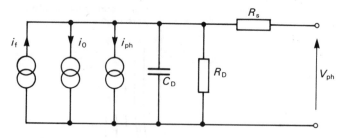

Fig. 4.7　Equivalent circuit of junction photodiode.

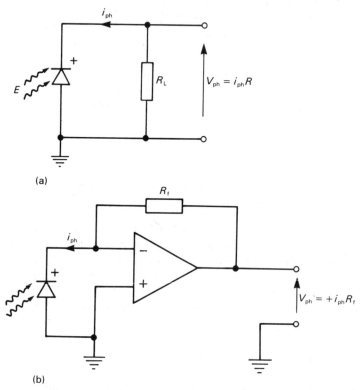

Fig. 4.8　Photodiodes connected in photovoltaic mode. (a) Simple circuit: logarithmic
response. (b) Linear response.

parallel with the voltage axis (line A in Fig. 4.6). The output response is clearly
non-linear, but does have good sensitivity and reasonable outputs can be obtained
at low irradiances. The non-linear response can be overcome to some extent by
using a low value of load resistor of around 1 kΩ, which has the effect of shifting
the load line until it is almost parallel with the current axis (Fig. 4.6 Line B). This
clearly allows a wider linear dynamic range, although the sensitivity at low
irradiance suffers. An alternative approach to provide a low load resistance is to
utilize the 'virtual earth' properties of an *op-amp* by connecting the diode directly
across its input terminals (Fig. 4.8b). This circuit maintains the low-noise characteristic of the PV mode but response times are still poor.

The general-purpose diode described in the data sheet of Table 4.2 is to be used as the basis of a simple light detector. Design photovoltaic circuits which will give (i) logarithmic and (ii) linear responses over as wide a dynamic range as possible.

Solution: The simple circuit of Fig. 4.8a will give a logarithmic response if $R_L \gg R_D$ and a linear response if $R_L \ll R_D$.

Since the data sheet gives us no indication of the value of R_D, we have to assume that if we choose as high a value of R_L as possible, the load line will be nearly parallel to the voltage axis and so ensure a logarithmic response. Hence, examination of the current-voltage curve shows that something of the order of 200 MΩ should suffice.

To ensure a linear response we need R_L to be low such that the load line is nearly parallel to the current axis, hence values of less than 1000 Ω will be acceptable.

For PC-mode operation, either of the circuits of Fig. 4.9 is suitable. In the basic 'diode-resistor' chain, the load line is determined by choice of bias potential and load resistor. To ensure linear response over as wide a dynamic range as possible, R_L should be as low as possible when compared with the dynamic resistance of the diode (line C in Fig. 4.6). When an op-amp with a high input resistance is used, such as a FET-amp, R_L can be as high as 1 MΩ.

In the *op-amp* based circuit (Fig. 4.9b) the output current is fed directly to the negative input of the amplifier. The circuit then functions as a *transconductance amplifier*. Typical feedback resistor values are usually around 1 MΩ, although this depends on the magnitude of the required output signal. Resistor R_S is included to balance the amplifier offset current and can be omitted for high input photocurrents.

See Horrocks, D.H. *Feedback Circuits and Op. Amps* (Van Nostrand Reinhold, 1983).

Using the high-speed photodiode (BPX 65) given in the data sheet (Table 4.2), design a photoconductive detector circuit to produce a 5 V output on a digital voltmeter (DVM) for a 5 mW input from the HeNe laser.

Solution: The circuit shown in Fig. 4.8b should be adequate for this purpose. The response of the BPX 65 at λ = 633 nm is, from the data, 0.4 A W^{-1}.

For a 5 mW laser beam the induced photocurrent will be

$$i_{ph} = 0.4 \text{ A W}^{-1} + 5 \text{ mW}$$
$$= 2 \text{ mA}$$

Since the op-amp circuit in this configuration will act as a transconductance amplifier, its output is given by,

$$V_{ph} = -i_{ph}R_f$$

and therefore,

$$R_f = V_{ph}/i_{ph}$$
$$= 5 \text{ V}/2 \text{ mA}$$
$$= 2.5 \text{ k}\Omega$$

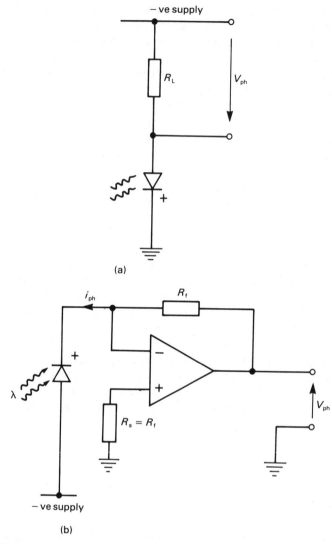

Fig. 4.9 Photodiode connected in photoconductive mode. (a) Simple circuit. (b) Op-Amp based circuit.

Photomultipliers

Some materials, notably alkali metals like sodium, lithium, caesium and their alloys, liberate electrons from their surface when illuminated by an external source of light. This is the well-known phenomenon of photoelectric emission and is the basis of operation of photoemissive detectors such as the vacuum phototube and the photomultiplier. In modern optoelectronics, though, it is the photomultiplier which is more commonly used and is the one we shall describe here.

Electrons are ejected from the photosensitive surface when the incident photons have enough energy to free the electron from its bond and remove it from the material (Fig. 4.10). This energy corresponds to the energy difference between the top of the valence band and the ionization level of the material and so represents

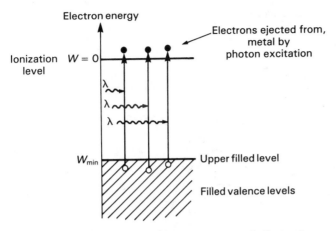

Fig. 4.10 Photoelectric emission from alkali metal.

the minimum energy required to eject an electron from the photosensitive surface. The consequence of this behaviour is that there is a threshold, or cut-off, wavelength above which no photoelectrons will be emitted. This condition may be expressed as,

$$W_{min} = hc/\lambda_c \qquad (4.13)$$

where λ_c is the *cut-off* wavelength.

Photomultipliers normally consist of a photosensitive surface, known as the *photocathode*, a series of secondary electrodes, called *dynodes* and a collector of photoelectrons, the *anode* (Fig. 4.11). Light impinging on the photocathode ejects electrons from the surface. The electric potential between the cathode and anode accelerates the photoelectrons towards the chain of dynodes, each of which is at a slightly higher potential than the one before it. As each electron strikes a dynode, a number of secondary electrons are produced, each of which produces further electrons at successive dynodes. The overall electron gain can be as high at 10^6.

The construction shown in Fig. 4.11 is the common *box and grid* structure. Other configurations, such as the *venetian blind, linearly* or *circularly* focused, offer improvements in size or performance. The venetian blind and box and grid structures, for example, are inexpensive and useful for large-area devices, whereas the focused structures offer higher collection efficiencies, signal modulation frequencies and rise times as low as a nanosecond.

Materials in common use as photocathodes are classified by the so-called

Manufacturers' literature, such as that by EMI, Centronic and Hamamatsu, contain discussions of other configurations and relative performance.

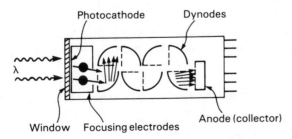

Fig. 4.11 Basic 'box and grid' photomultiplier.

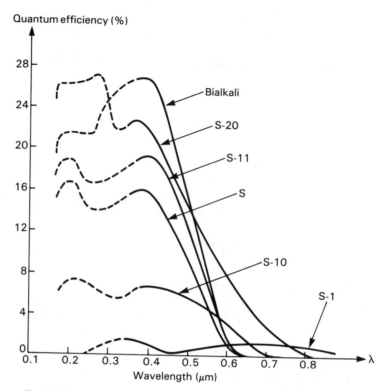

Fig. 4.12 Quantum efficiency for various photocathode types.

S-system. Monoalkali photocathodes, such as S-1 and S-11, cover a wide spectral range at low cost, but are usually inefficient and noisy by comparison with solid-state detectors. Bialkali materials, like K-Cs or Na-K, offer low noise at moderate efficiencies, whereas the trialkalis, such as S-20, offer the ultimate in performance, but obviously at a price. Quantum efficiencies of the more common photocathode materials are shown in fig. 4.12.

In the evaluation of PM performance, the important parameters to consider are the responsivity, gain, tube voltage and dark current. Two figures are usually given to describe PM responsivity: one, the cathode responsivity, refers to the photo-current generated at the photocathode; and the other, the anode responsivity, refers to the amplified photocurrent after passing through the dynode chain. The overall gain of a photomultiplier is expressed as the ratio of its anode responsivity to the cathode responsivity.

Worked Example 4.5 A particular photocathode has a responsivity of 90 μA lm^{-1}. If it were used in a PM with a dynode gain of 3.0×10^6, what is the overall sensitivity of the instrument?

Solution: The overall responsivity, or anode responsivity, is given by

$$S_A = \text{(cathode responsivity)} \times \text{(dynode gain)}$$
$$= 90 \times 10^{-6} \, \text{A lm}^{-1} \times 3 \times 10^6$$
$$= 270 \, \text{A lm}^{-1}$$

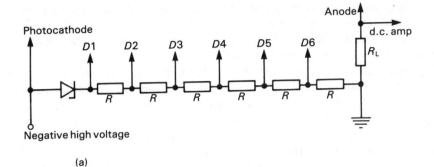

Fig. 4.13 Dynode chain configuration. (a) Continuous operation.
(b) Pulse operation. (c) Ground cathode.

Dynode Biasing Chains

To use PM tubes, each consecutive dynode has to be biased to successively higher potentials. This ensures that each dynode presents an attractive force to the electrons ejected from the preceding dynode. A simple but effective means of accomplishing this is with the potential-divider circuits shown in Fig. 4.13, which split the cathode-anode potential in nearly equal amounts between dynodes.

In designing such circuits, one or two important points should be borne in mind. Photomultipliers are normally designed to operate with a specific voltage, typically 150 V, between the cathode and first dynode: a constant potential is best maintained using a zener diode of the required potential between the cathode and D1. Secondly, to preserve linearity and provide as wide a dynamic range as possible, the current through the dynode chain should be around 10 to 100 times greater than the average anode current. Since anode currents of 100 μA are representative, we need no more than 1 to 10 mA through the chains. For cathode-anode potentials of 1000 to 2000 V, 100 kΩ resistors are good choices.

The basic circuit of Fig. 4.13a is the standard chain for measurement of continuous light levels. If the output is fed to a high-impedance measuring instrument such as a d.c. amplifier or a galvanometer, the load resistor can be omitted. For pulsed operation, the peak anode current can be many times greater than the average anode current, which in turn can lead to saturation and non-linearity of output. To overcome this, the potential between the last two or three dynodes is increased by increasing the corresponding resistor values (Fig. 4.13b). Another variation (Fig. 4.13c) is often employed in pulse counting applications where it is desirable to reduce noise generated by electrons colliding with other electrons in the glass envelope. This can be eliminated by reversing the potentials of the cathode and anode: a situation known as *ground cathode*. The capacitors employed across the last few dynodes help to maintain the required potentials at a constant value.

Thermal Detectors

Unlike all the preceding detectors which respond directly to photon collisions with electrons, thermal detectors respond to the heat content of the incident radiation. On irradiation, light is absorbed, the detector material increases in temperature producing a change in one or more of its electrical properties. Generally, because heat energy is the function causing the response, the spectral range is broad, though this can sometimes be limited by the detector window. Rise times are usually shorter than for photon detectors, although the pyroelectric detector is a notable exception, and can be as long as several seconds. Sensitivities, also, tend to be lower, but the devices can normally withstand much higher irradiances.

Thermocouples and thermopiles are probably the most commonly used thermal detectors and utilize the principle of the thermoelectric effect. A thermocouple is formed when two dissimilar metals or semiconductors are joined together at both ends. If one junction is heated with respect to the other, a potential difference is created across it resulting in current flow around the circuit. The other junction, called the *cold junction*, is kept at a constant reference temperature. The size of the voltage signal generated by one thermocouple is relatively small. Sensitivity, therefore, is enhanced by connecting several thermocouples together in series, so that the total developed potential is the sum of the individual potentials. This is the *thermopile* which is more commonly used in light detection than the basic thermocouple.

Two types of absorbing material are used in thermopiles: surface absorbers and volume absorbers. Surface absorbers, which absorb light over a wide wavelength range, are useful for all continuous power measurements but will not withstand the high peak powers generated by some pulsed lasers. Volume absorbers employ a glassy material which is partially transparent to radiation of a particular wavelength range. These detectors need to be chosen with particular reference to the wavelengths to be monitored, but even so their response is flat over a much wider range than photon detectors. They are particularly suitable for monitoring pulsed-laser flux and energy.

Pyroelectric Detectors

Certain crystalline materials exhibit a phenomenon known as *ferroelectricity* by which the existence of an internal electric field induces a charge distribution on opposite faces of the crystal. The origin of this field arises from the small permanent electric field associated with the individual atoms or molecules which comprise the crystal. When the crystal is well below a certain temperature, the *Curie temperature*, the individual fields align in the same direction to produce a net internal field parallel to the crystal axis. As the crystal is heated, thermal agitation of the molecules disrupts the alignment of the field and with it, the charge distribution on the crystal faces until, at the Curie temperature, alignment is totally destroyed.

A useful reference text for this chapter is Anderson, J.C., Leaver, K.D. Rawlings, R.D. and Alexander, J.M. *Materials Science* (Van Nostrand Reinhold, 1985).

Typical ferroelectric crystals, such as lithium tantalate or lead zirconate, form the sensitive element in pyroelectric detectors. The crystal faces are coated with a transparent material, permitting transmission of incident radiation, and which also acts as electrodes. Incident radiation produces local heating which changes the surface charge in the crystal and, hence, alters any stored charge in the electrodes. It is this change in the stored charge which is detected as a current flow in the external circuitry. Because it is a change in stored charge which is measured, pyroelectric detectors can only be used with continuous light beams if an external beam chopper is used. The sensitivity and spectral response are dependent on the chopping frequency. Unlike other thermal detectors, response times can be as short as a few nanoseconds.

The Bolometer

The bolometer is the last thermal detector we shall look at and is a useful, general-purpose detector. The general principle of the bolometer is that a sensing element such as a coil of metal wire or metal strip is heated by the incident radiation resulting in a change in its resistance. The change in resistance is measured by the external circuitry.

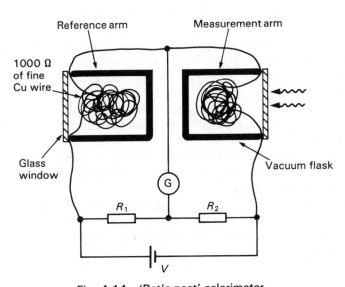

Fig. 4.14   'Rat's nest' calorimeter.

Other wire such as platinum or
nickel could obviously be used
but are more expensive.

See Baker, R.M. *Electronics*
Vol. 36, pages 36–8, 1983.

A simple bolometer can be easily and cheaply made from two or three hundred metres of copper wire and a small vacuum flask and will give a reasonable measurement of pulsed-laser energies. Cut a length of fine copper wire so that its resistance is about 1000 Ω and bundle the wire into the vacuum flask. Make sure the free ends of the wire are hanging out of the open end of the flask and then seal with a glass plate. We now have what is fondly known as a *Rat's Nest Calorimeter*. The 'nest' is used as one arm of a Wheatstone bridge circuit to produce the final instrument (Fig. 4.12). Although a mathematical relation can be derived linking input energy with resistance change, the best way to use the device is to calibrate it with known values of input energy. Performance can be improved by making a second, identical, rat's nest and using it, instead of a standard resistor, as the 'reference' arm. Used in this way the 'rat's nest' will give a reasonable measurement of pulsed-laser output powers.

Summary

In this chapter we have introduced the fundamentals of photodetectors, how they respond to incident radiation and how this response is monitored to give a measurement of light flux. We have discussed the language by which photodetectors can be compared in terms of their performance and have described the operation of the most useful photodetectors such as junction photodiodes, LDRs, photomultipliers and thermal detectors. In addition we have looked at some typical detector circuitry.

Problems

4.1 An LDR (type ORP 12) connected to a 12 V relay is to form the basis of a light-activated alarm. If the relay needs a minimum current of 100 mA to prevent it from triggering the alarm, what is the minimum luminous flux which must be present on the detector? Assume the light source to have a peak wavelength of 550 nm.

4.2 A measurement of both the output energy and temporal profile of pulsed Nd-YAG laser is needed. Using the data in Tables 4.1–4.4 choose a suitable detector or detectors. The maximum output energy of the laser is 250 mJ over a pulse duration of 15 ns.

4.3 If in the previous problem, simultaneous measurement of both peak power and pulse duration are needed how does this influence your selection of detector/s?

4.4 If a radiant flux of 50 μW falls directly on a photodetector with a sensitivity of 0.5 A W^{-1} at the peak emission wavelength of the source, calculate the current generated in the detector.

4.5 Using the large area photodiode shown in Table 4.2 and basing your design on the photovoltaic circuit shown with a feedback resistance of 10 kΩ, estimate the output voltage obtained when the diode is irradiated by a 1 mW cw HeNe laser beam of 1 cm² diameter.

Table 4.2
Selected junction photodetectors

Type	General purpose	High speed (BP × 65)	Large area
Manufacturer	Radiospares	Radiospares	Radiospares
Junction	Si, pn	Si, pin	Si, pn
Responsivity	$0.7 \ \mu$A mW^{-1} cm^{-2} (0 V)	0.55 A W^{-1} (850 nm)	0.2 A W^{-1} (450 nm)
			0.35 A W^{-1} (633 nm)
			0.5 A W^{-1} (900 nm)
			0.15 A W^{-1} (1064 nm)
Active area	0.85 × 0.85 mm	1 mm^2	1 cm^2
Capacitance	12 pF (-10 V)	3.5 pF (-20 V)	1500 pF (0 V)
Dark current	1.4 nA (-20 V)	1 nA (-20 V)	$0.5 \ \mu$A (-10 V)
Response time	250 ns (10–90%)	0.5 ns (-20 V)	$0.5 \ \mu$s (0 V)
Reverse voltage (max)	80 V	50 V	100 V
Forward current (max)	100 mA	10 mA (dc)	Limited by bias
Operating temperature	0 to +70°C	-25 to +70°C	-55 to +70°C
Power dissipation	200 mW	250 mW	100 mW

Table 4.3
Selected photomultiplier tubes

Type Number	4181	4162	4272	9597	9664
Manufacturer	Centronic	Centronic	Centronic	EMI	EMI
No. of stages	11	9	10	14	9
Overall responsivity	200 A lm^{-2}	50 A lm^{-1}	50 A lm^{-1}	5000 A lm^{-1}	20 A lm^{-1}
Cathode responsivity	66 μA lm^{-1}	190 μA lm^{-1}	83 μA lm^{-1}	140 μA lm^{-1}	20 μA lm^{-1}
Anode/cathode voltage	950 V	950 V	1000 V	2300 V	750 V
Rise time	15 ns	15 ns	6 ns	2 ns	2 ns
Dark current	1.6 nA	0.08 nA	0.1 nA	300 nA	2 nA
Size (envelope diameter)	29 mm	29 mm	51 mm	52 mm	30 mm
Structure	Box and Grid	Box and Grid	Venetian blind	Linear focused	Compact focussed
Photocathode	S-11	Bialkali	Bialkali	S-20	S-10

Table 4.4
Selected thermal detectors

	Thermal detectors		
Type	Thermopile	Thermopile	Pyroelectric
Absorber	Surface	Volume	—
Spectral range	190 nm–35 μm	375–1300 nm	250 nm–12 μm
Aperture size	25 mm	25 mm	10 mm
Power range	100 μW–75 W	100 μW–75 W	10 μW–10 W
Power density CW (max)	200 W cm^{-2}	30 W cm^{-2}	200 W cm^{-2}
Power density, pulse (max)	1 MW cm^{-2}	10^{11} W cm^{-2}	—
Single pulse (max)	—	15 J	—
Responsivity	100 mV W^{-1}	3 mV J^{-1}	
Responsivity	12 s	15 s	0.1 s
Manufacturer	Photon Control	Photon Control	Laser Precision

5 Radiometry and Light Coupling

Objectives
- [ ] To revise concepts of image formation in optical systems
- [ ] To introduce principles of radiometry
- [ ] To discuss the spatial profile of some common light sources
- [ ] To introduce concepts of light coupling between source and detector
- [ ] To outline principles of light transmission through optical fibres

Having looked in detail at the operation of some sources and detectors commonly used in optoelectronics, it now rests with us to bring these components together in the formation of an optoelectronic system. Our prime task in this chapter is to discuss what we might loosely call 'light economics', or how to optimize the coupling of light between source and detector. The measurement of optical radiation and its transfer through optical systems is known as *radiometry* and it is here that the foundations of this chapter lie.

Ray Optics

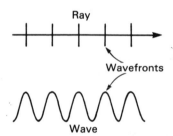

Refer to Chapter 2 for a discussion of electromagnetic waves.

In Chapter 2 we saw that to describe the behaviour of light it was necessary to invoke two distinct models, namely, the wave and photon models. When designing optical systems, however, we can simplify our approach by neglecting the wave and photon character of light and considering only the direction of propagation of the wavefront through the system.

The direction of travel of an optical wave is defined in terms of straight lines, known as *rays*, which propagate at right angles to the wavefront. In other words, ray propagation is mutually perpendicular to the electric and magnetic vectors. This approach forms the basis of *ray optics* or *geometric optics*. A further simplification of the design process can be invoked, depending on the angle which the rays make with the *optical axis*. The optical axis is the central ray through the system. By considering only those rays which propagate at angles close to the optical axis, the so-called *paraxial* region, we arrive at a series of simple equations which form the basis of our design calculations. In fact the paraxial approach is often extended out of its regime of strict validity to cover rays which propagate at large angles to the optical axis with surprisingly good results.

The Lens Equation

See for example Longhurst R.S. *Geometrical and Physical Optics* (Longman, 1974).

For our purposes, a simple consideration of geometric optics will enable us to design a wide variety of optical systems. A simple lens system is shown in Fig. 5.1, where a lens of focal length f, images a source of height h to size h'. If the source-to-lens distance is u and the image is formed at a distance v behind the lens then the following relationship holds:

$$1/u + 1/v = 1/f \tag{5.1}$$

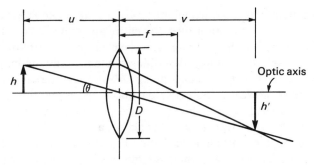

Fig. 5.1   Simple image formation.

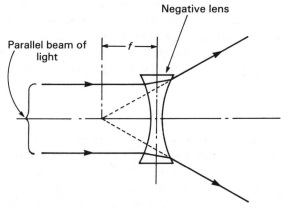

Fig. 5.2   Divergence of parallel beam of light using negative lens.

This is the well-known *lens equation*. The magnification is given by,

$$m = h'/h = u/v \qquad\qquad (5.2)$$

We should bear in mind that these relations hold for *thin* lenses and for points *close to the optic axis*. This is an example of the paraxial approximation.

The system shown in Fig. 5.1 makes use of a lens which forms a *real* image of the source on the opposite side of the lens to that on which the light is incident. The light *converges* to a focus. Such a lens is said to be a *positive* lens. *Negative* lenses form an apparent, or *virtual*, image of the source on the same side of the lens as the incident light. Such lenses *diverge* the incident light. The use of a negative lens to diverge a parallel beam of light is shown in Fig. 5.2.

We shall see the use of negative and positive lenses in Chapter 6.

A source of 1 mm² active area is positioned 50 mm from a thin, convex lens of 100 mm focal length. Calculate the position and size of the image so formed.   *[100 mm, 4 mm²]*

Exercise 5.1

Radiometry and Photometry

To assess the performance of an optical system we need more than an equation which tells only where an image is formed: we also need some means of *quantifying* the amount of light passing through the system. *Radiometry* deals with the measurement of optical radiation over the entire spectrum from ultra-violet to infra-red. *Photometry*, on the other hand, is the measurement of visible light as it appears to the human eye. Although describing fundamentally similar parameters, namely *flux, energy, intensity, radiance* and *irradiance*, a different system of units is used to distinguish between radiometric and photometric quantities. A particular algebraic symbol is used to describe the optical parameter, be it radiometric or photometric, with the subscript 'e' added to denote a radiometric quantity and the subscript 'v' to denote a photometric quantity. For example, radiant flux, a radiometric quantity, is denoted by Q_e, whereas luminous flux, a photometric quantity, is denoted by Q_v. The principal radiometric and photometric parameters are summarized in Table 5.1.

An important point to note is that the radiometric and photometric parameters as we have defined them here refer to the total amount of radiation emitted summed over all wavelengths. Sometimes we may want to define a radiometric parameter over a reduced frequency or wavelength interval. For example, *spectral radiance* is the flux per unit area per unit solid angle *per unit frequency interval* and denoted as $L_e(\lambda)$.

Photopic Response

We have already made passing reference to this fact in Chapter 2.

The distinction between radiometry and photometry is necessary because the eye responds unequally to the various wavelengths of light. Two monochromatic light sources of wavelengths λ_1 and λ_2 may have the same radiant flux but different values of luminous flux. A green-light-emitting diode, for example, will appear brighter than a red one of the same radiance. Radiant flux and luminous flux are related via the *photopic response* curve of the human eye (Fig. 5.3). At the peak of the curve, 550 nm,

$$1 \text{ W} \equiv 674 \text{ lm} \tag{5.3}$$

Worked Example 5.1   A particular light-emitting diode made from GaAsP emits 25 μW of flux at its peak wavelength of 650 nm (red). Another diode made from GaP emits the same flux at its peak of 550 nm (green). From the photopic response curve, the photopic sensitivity of GaAsP at 650 nm is 0.107, whereas, for GaP at 550 nm it is 0.967. The luminous flux of the GaAsP diode is therefore

$$\Phi_v = 25 \times 10^{-6} \text{ W} \times 674 \text{ lm W}^{-1} \times 0.107$$
$$= 1.8 \text{ mlm}$$

The corresponding flux for the GaP diode is

$$\Phi_v = 25 \times 10^{-6} \text{ W} \times 674 \text{ lm W}^{-1} \times 0.967$$
$$= 16.3 \text{ mlm}$$

For the GaAsP to produce the same visual effect as the GaP diode, its radiant flux would have to be increased by the ratio of the luminous flux of the GaAsP diode to that of the GaP diode. Hence, the required radiant output is

$$\Phi_e = 25 \times 10^{-6} \text{ W} \times 16.3 \text{ mlm}/1.8 \text{ mlm}$$
$$\therefore \Phi_e = 214 \text{ } \mu\text{W}$$

TABLE 5.1: RADIOMETRIC AND PHOTOMETRIC QUANTITIES

QUANTITY	DESCRIPTION	SYMBOL [Radiometric] [Photometric]	DEFINITION	UNITS [Radiometric] [Photometric]	GEOMETRIC RELATIONSHIP
Energy	Amount of light emitted	Q_e Q_v		J J	
Flux (Radiant power)	Rate of energy emitted or transferred through space	Φ_e Φ_v	$\Phi = \dfrac{dQ}{dt}$	W lm	
Intensity	Flux emitted from source per unit solid angle	I_e I_v	$I = \dfrac{d\Phi}{d\omega}$	W sr^{-1} lm sr^{-1} (cd)	$\omega = \dfrac{A}{r^2}$
Emittance	Flux emitted per unit area of source	M_e M_v	$M = \dfrac{d\Phi}{dA}$	W m^{-2} lm m^{-2} (lx)	
Radiance Luminance	Flux emitted per unit area of source, per unit solid angle, at angle θ to surface normal	L_e L_v	$L = \dfrac{d\Phi}{dA\, d\omega \cos\theta}$	W m^{-2} sr^{-1} lm m^{-2} sr^{-1} (cd m^{-2})	
Irradiance Illuminance	Flux arriving per unit area of irradiated surface at angle θ to surface normal	E_e E_v	$E = \dfrac{d\Phi}{dA \cos\theta}$	W m^{-2} lm m^{-2} (lx)	

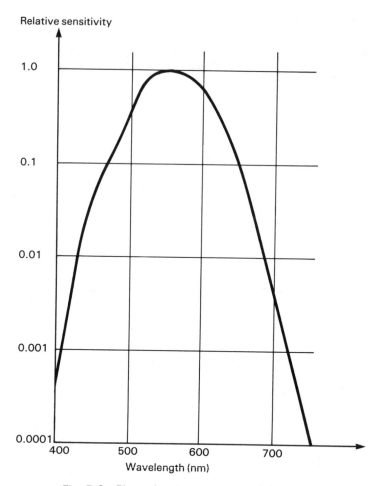

Relative sensitivity

1.0

0.1

0.01

0.001

0.0001

400   500   600   700

Wavelength (nm)

Fig. 5.3 Photopic response curve of the eye.

Spatial Radiation Profile

Successful optical system design requires that we know something of the *spatial* character of the radiation emitted from the source. Gas lasers, for example, emit light in a narrow pencil of a few milliradians divergence, whereas an LED or laser diode radiates into an elliptical cone of some 30° or 40° half angle and incandescent lamps possesses an almost spherical radiation profile. Some other common profiles are the *line radiator* and the *Lambertian radiator*.

The calculation of the basic radiometric parameters and their relationship to the spatial profile of a source is best seen by way of one or two examples. Firstly, lets consider the ideal spherical radiator: the *isotropic point source*. By definition an isotropic point source is one which emits light uniformly in all directions in space. The almost spherical distribution of light around a standard incandescent lamp provides us with a reasonable approximation to the ideal point source.

Worked Example 5.2 Calculate the intensity and radiance of a 100 W incadescent light bulb and estimate the flux received at a flat surface of 0.1 m² area positioned normal to the light bulb and 1 m from it.

Solution: Assuming that we can approximate the radiation profile of the lamp by that of a spherical point source, we can estimate the radiant intensity emitted by the bulb from the relation

(See Table 5.1.)

$$I = d\Phi/d\omega \tag{5.4}$$

where $d\omega$ is an element of solid angle.

Solid angle is defined in terms of the angle subtended at the centre of a sphere of radius r by an element dA_i of the surface area of the sphere, given by $d\omega = dA_i/r^2$. Since the surface area of a sphere is $4\pi r^2$, the solid angle subtended by the surface is

$$\Omega = 4\pi r^2/r^2 = 4\pi \quad \text{steradians (sr)} \tag{5.5}$$

For an isotropic point source the total flux emitted is radiated into a 4π solid angle and, hence, its radiant intensity is,

$$I = 100 \text{ W}/4\pi \text{ sr}$$
$$\sim 8 \text{ W sr}^{-1}$$

Now to calculate the flux received at the irradiated surface we need to calculate, first, the solid angle, $d\omega$ subtended at the source by an element of the sphere's surface area, dA_i. Hence for our situation we have,

$$d\omega = dA_i/r^2$$
$$= 0.1 \text{ m}^2/(1 \text{ m})^2$$
$$= 0.1 \text{ sr} \tag{5.6}$$

We have assumed here that for small values of ω the surface of a small portion of the sphere is flat. Hence, the flux received at the surface, is,

$$\Phi = I d\omega$$
$$= 8 \text{ W sr}^{-1} \times 0.1 \text{ sr}$$
$$= 800 \text{ mW}$$

The *irradiance* at the surface is defined as the flux per unit area received, and hence

$$E = d\Phi/dA_i$$
$$= 800 \text{ mW}/0.1 \text{ m}^2$$
$$= 8 \text{ W m}^{-2} \tag{5.7}$$

Another useful source profile is that corresponding to the line radiator. In this instance, the source emits uniformly in a cylindrical pattern around the axis of the radiator (Fig. 5.4). A good practical approximation to this profile is provided by a linear flashtube of the type used in electronic flashguns or solid-state lasers.

The flux emitted by a cylindrical line radiator of length l and normal intensity I_o is described by

Exercise 5.2

$$\Phi = \pi^2 I_o \tag{5.8}$$

and its radiance at a point r perpendicular to the axis of the source is

$$L = I_o/2rl \tag{5.9}$$

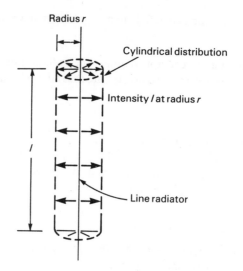

Fig. 5.4 Cylindrical distribution of light around line radiator.

We will not derive these relations, but merely state them. The similarity between the equations quoted here and corresponding relations relating to electric fields should be noted.

Estimate the flux emitted and radiance of a 35 mm × 3 mm diameter flashtube of 100 W sr^{-1} normal intensity.

[987 W, 952 kW m^{-2} sr^{-1}]

One final source profile which is commonly encountered is the *Lambertian radiator*. A Lambertian source is one which is subject to the *Lambert cosine law* (Fig. 5.5) where its intensity in any direction at θ to the normal is given by,

$$I = I_0 \cos\theta \qquad (5.10)$$

In a Lambertian source the radiance given by

$$L = d\Phi/\pi dA_e \qquad (5.11)$$

is *constant* regardless of the angle from which it is viewed. This behaviour is closely approximated to by illuminating a ground-glass screen from behind. The radiation pattern emitted by some LEDs is similar to that of a Lambertian source.

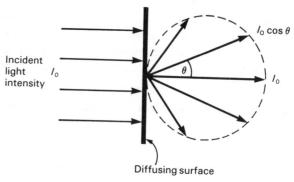

Fig. 5.5 Radiation profile from Lambertian surface.

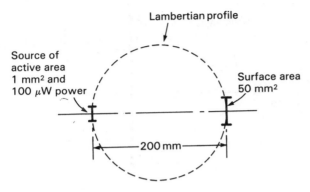

Fig. 5.6 Worked Example 5.3.

Worked Example 5.3

An LED of 1 mm² active area emits 100 μW of radiation in an approximately Lambertian profile. Estimate the irradiance at a surface of area 50 mm² placed normal to the source and 200 mm from it.

Solution: The system geometry is shown in Fig. 5.6. The total flux emitted from a Lambertian source is given by

$$\Phi = \pi L dA_e \tag{5.12}$$

From eqn (5.12) we can calculate the radiance of the source as

$$\begin{aligned} L &= \Phi/\pi \, dA_e \\ &= 100 \, \mu\text{W}/(\pi \times 1 \text{ mm}^2) \\ &\simeq 32 \text{ W m}^{-2} \text{ sr}^{-1} \end{aligned}$$

The solid angle subtended at the source by the surface is

$$\begin{aligned} d\omega &\simeq dA_i/r^2 \\ &= 50 \text{ mm}^2/(200 \text{ mm})^2 \\ &= 1.25 \text{ msr} \end{aligned}$$

Hence, the flux received by the surface is

$$\Phi = 32 \text{ W mm}^{-2} \text{ sr}^{-1} \times 1 \text{ mm}^2 \times 1.25 \text{ msr}$$
$$\therefore \Phi = 40 \text{ nW}$$

Coupling of Light

In an optoelectronic system, our simple aim is to transfer radiant energy from the source to the detector. We can do this in two ways, either by *direct tranfer* of light in which there are no optical components between source and receiver, or by *indirect transfer* in which a system of lenses or light guides are used. In either case, our basic design criterion is to aim for *maximum transfer of flux* between source and detector and so ensure the greatest sensitivity of our system. In so doing the fundamental principle we must keep in mind is that, if lens and atmospheric losses are taken into account, *the radiance throughout the system is constant.*

See Longhurst, R.S. *Physical and Geometric Optics* (Longman, 1974).

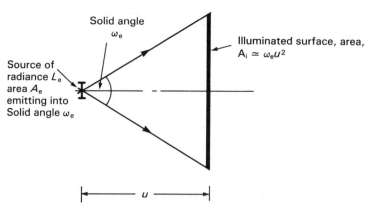

Fig. 5.7   Direct irradiation of a surface.

Direct Illumination

In direct illumination, no lens comes between the detector or irradiated surface. The total area illuminated is dependent on the spatial profile of the source and its location. The flux which falls on the surface is therefore proportional to its area (Fig. 5.7). Hence, the flux received at the surface when irradiated by a source of radiance L_e is

$$\Phi_i = L_e A_e \omega_e$$
$$= L_e A_e A_i / u^2 \tag{5.13}$$

where A_e is the radiant area of the source and ω_e is the solid angle subtended at the source by a surface of area A_i placed at a distance u from the source.

Once again we assume paraxial conditions.

To receive the maximum available flux the detector area should equal that of the irradiated area of the surface. When a detector of area A_i is used to collect the light then eqn (5.13) will give the received power.

Effect of Mismatching

If a detector of area A_d (Fig. 5.8) is placed at the irradiated surface the flux collected by the detector is given by,

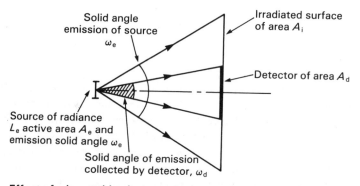

Fig. 5.8   Effect of mismatching between source emission profile and collection angle of detector.

$$\Phi_d = L_e A_e \omega_d \qquad (5.14)$$

where ω_d is the solid angle subtended at the source by the detector. If, as is generally the case, the irradiated and detector areas are unequal then the ratio of detected to transmitted power is

$$\Phi_d/\Phi_i = L_e A_e \omega_d / L_e A_e \omega_e$$
$$= \omega_d/\omega_e \qquad (5.15)$$

Since $\omega_d = A_d/u^2$ and $\omega_e = A_i/u^2$, we have,

$$\Phi_d/\Phi_i = A_d/A_i$$
$$= \eta_c \qquad (5.16)$$

where η_c is defined as the *collection efficiency* of the detector. A simple example will illustrate the use of collection efficiency.

This is sometimes erroneously called the optical transfer function (OTF). In optics, OTF has a very specific meaning, which is outside the scope of this text.

Worked Example 5.4

If a source of 5 mW output flux irradiates a surface area of 10 cm², what is the flux received by a 1 cm² detector placed at the surface, normal to the source and on the optical axis?

Solution: In this example we assume that to detect the maximum flux, the source and detector are on the same optical axis and their surfaces are normal to it, as in Fig. 5.9. The collection efficiency of the detector is therefore

$$\eta_c = A_d/A_i$$
$$= 1 \text{ cm}^2/10 \text{ cm}^2$$
$$= 0.1$$

In optics, angles are measured with respect to the perpendicular through any surface, known as the 'normal'. Hence, 'normal to' and axis means 'at right angles to'.

The detected flux is thus

$$\Phi_d = 0.1 \times 5 \times 10^{-3} \text{ W}$$
$$\therefore \Phi_d = 0.5 \text{ mW}$$

Herein lies the main disadvantage of direct illumination: it is wasteful of energy and does not fulfill the criterion for maximum flux transfer. In this situation the detector will respond to the flux per unit area, or the *irradiance*.

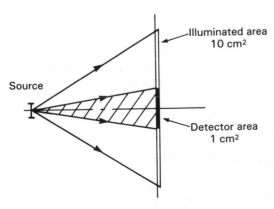

Fig. 5.9 Worked Example 5.4.

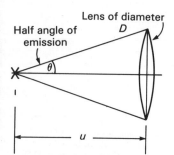

Lens of diameter D

Half angle of emission

θ

Numerical aperture of lens

The refractive index of air is ~1.

Note $A_i = \pi D^2/4$.

The Numerical Aperture

From the previous few sections, we can see that the fundamental parameter which allows us to determine the amount of light incident on a surface, is the *collection cone* of light delineated by the solid angle subtended at the source by the surface area. In most of the situations with which we are concerned, the irradiated surface will be flat and not curved as is implied in the solid-angle relation. In the paraxial regime, however, by assuming that ω is small, we can overlook this disparity.

An alternative, and slightly simpler approach to calculating collection efficiencies, is given using the concept of numerical aperture (NA) of a lens instead of the solid angle. The numerical aperture is defined as,

$$NA = n \sin \theta \tag{5.17}$$

where n is the refractive index of the medium and θ is the half angle between the optical axis and the widest-angle ray which the lens will accept. Yet again invoking the paraxial approximation, that is, when θ is small, we have

$$\sin \theta \simeq \tan \theta$$

and therefore

$$NA \simeq \tan \theta$$
$$NA \simeq D/2u \tag{5.18}$$

where D is the diameter of the illuminated area. Thus if the illuminated area is A_i, the numerical aperture will, therefore be given as

$$NA_e = (A_i/\pi)^{\frac{1}{2}}/u$$

or, the square of the numerical aperture is,

$$(NA_e)^2 \sim A_i/\pi u^2 \tag{5.19}$$

Returning now to the definition of collection efficiency, if we have an irradiated surface of area A_i and a detector of area A_d the collection efficiency of the detector is given by,

$$\eta_d = A_d/A_i = (NA_d)^2/(NA_e)^2 \tag{5.20}$$

This simple relation enables us to design a wide variety of optical systems to a first approximation.

We can see how the numerical aperture approach works by considering the following example.

Worked Example 5.5

Note this is the same example as Worked Example 5.4, but stated slightly differently.

If a source emits 5 mW into a half angle of 30°, calculate the flux detected by a 1 cm² detector at 100 mm from the source.

Solution: The numerical aperture of the source is

$$NA_e = \sin 30° = 0.5$$

The maximum irradiated area is, therefore, from eqn (5.19)

$$A_i = \pi u^2 (NA_s)^2$$
$$= 3.14 \times (100 \times 10^{-3} \, m)^2 \times (0.5)^2$$
$$= 10 \, cm^2$$

96

Since the detector area is only 1 cm², the flux detected by it will be,

$$\Phi_d = 5 \times 10^{-3}\,\text{W} \times 1\,\text{cm}^2/10\,\text{cm}^2$$
$$\therefore\ \Phi_d = 0.5\,\text{mW}.$$

In the special case when the source is exactly one focal length away from the lens, that is when $u = f$, then from eqn (5.18),

$$\text{NA} = D/2f$$

and so,

$$f/D \sim 1/2\text{NA}.$$

The term f/d is the photographic *f-number* or *aperture* of a lens. We saw in Chapter 3 that the spot diameter to which a laser beam is focused is $d = 2f\Delta\theta$, and also that $\Delta\theta$ is given approximately by λ/D, hence,

$$d \simeq \lambda f/D$$

and

$$d \simeq \lambda/\text{NA}$$

This is the relation we used in Chapter 1 to calculate the spot diameter in a CD player.

Indirect Illumination

In indirect illumination, some means of collecting the emitted light from the source is employed. Usually this is via a lens, light guide or combination of both (Fig. 5.10). The diameter and focal length of the lens are chosen so as to collect as much of the light emitted from the source as possible. Too small a diameter and not all the light will be intercepted; too large a diameter and money is wasted on an over-specified component. The basic criterion that is employed is that of *matching the numerical aperture of the lens with that of the source.*

A procedure sometimes known as *matching f-numbers.*

In this case the collection efficiency of the lens is given as

$$n_1 = (\text{NA}_l)^2/(\text{NA}_e)^2$$

where NA_l is the numerical aperture of the collection lens. The collection efficiency of the detector is determined by the numerical aperture of the collection lens and that of the detector. Hence

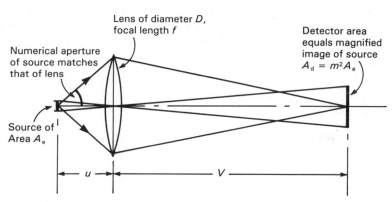

Fig. 5.10 Indirect irradiation showing optimum matching of numerical apertures.

$$\eta_d = (NA_d)^2/(NA_l)^2$$

The collection efficiency of the combined system will therefore be related to the product of all the collection efficiencies in the system. That is,

$$\eta_{sys} = \eta_l \times t_l \times \eta_d \tag{5.21}$$

or

$$\eta_{sys} = [(NA_l)^2/(NA_e)^2] \times t_l \times (NA_d)^2/(NA_l)^2 \tag{5.22}$$

For an ordinary uncoated lens t can be as low as 0.7 or 0.8; for a lens with an anti-reflection coating t will be in the 0.95–0.99 range.

where t_l is the transmission factor of the lens.

If possible, the detector area is chosen in conjunction with the focal length of the lens, so that the ratio of detector area to source area equals the magnification of the system, thereby, ensuring that all light collected by the lens is imaged onto the detector. In this case the collection efficiency of the system is given by

$$\eta_d = [(NA_l)^2/(NA_e)^2] \times t_l \tag{5.23}$$

Again let us illustrate the procedure by way of an example.

Worked Example 5.6

Design a system which will deliver 5 mW from a source of 30° full emission angle and 10 mm² area to a detector of 1 cm² area with a minimum of light loss.

Solution: Our first step in the design of the required system is to calculate the numerical aperture of the source. This is given by

$$NA_e = \sin\theta = \sin15° = 0.26$$

For optimum coupling, our chosen lens should also have a numerical aperture of this value, namely, 0.26.

Now we estimate the system magnification required to just fill the detector with light. In this case we have

$$m = \sqrt{(A_d/A_e)} = \sqrt{(1\text{ cm}^2/10\text{ mm}^2)} = \sqrt{10} \simeq 3$$

A magnification of less than three would still produce a measurement of the total power on the detector. Overfilling the detector with a magnification of more than three would yield the power per unit area, or irradiance.

Using the system magnification in conjunction with the lens equation (5.1) will yield the focal length of the lens f, the source-to-lens distance u and the lens-to-detector distance v. However, we have to make an initial choice of one of the above parameters, or instead the total distance between source and detector, $u + v$.

In this example, let us assume that source and lens can be no closer than 50 mm, this dictates that for a magnification of 3 the lens-to-detector distance must be 150 mm. Substituting these values for u and v into the lens equation gives

$$1/f = (1/50) + (1/150)$$

and hence

$$f \simeq 38\text{ mm}$$

The minimum lens diameter which will allow collection of all the source light is given by

$$NA \simeq D/2u$$

Thus,

$$D \simeq 2 \times 50 \times 10^{-3}\,\text{m} \times 0.26$$
$$D = 26\,\text{mm}$$

Hence for optimum coupling of light we need a lens of 38 mm focal length, 26 mm diameter placed 50 mm from the source and 150 mm from the detector.

In practice, unless we wish to have a lens specially made, we should choose the most suitable standard lens, say $f = 40$ mm and $D = 25$ mm and recalculate the corresponding imaging conditions. Another point to be considered is that when trying to form a high-resolution image of the source, because of aberrations in the lens, rays of light which pass through the extremities of the lens tend to degrade the image. This occurrence again leads us into choosing a lens diameter slightly higher than that calculated.

In many cases we will be concerned with a multi-lens system in which the light may collected by one lens, relayed to another and then imaged onto a detector. In this case the procedure is merely an extension of that described above: optimum matching is obtained when the output numerical aperture of the first lens equals the input numerical aperture of the second lens, and so on. This results in the following chain equation,

$$\begin{aligned}
\eta_{\text{sys}} &= \eta_1 t_1 \times \eta_2 t_2 \times \eta_d \\
&= [(NA_1)^2/(NA_c)^2] \times t_1 \times [(NA_2)^2/(NA_1)^2] \times t_2 \\
&\quad \times (NA_d)^2/(NA_2)^2
\end{aligned} \tag{5.24}$$

and so on, for any number of components.

The following exercise will give you the opportunity of trying an example for yourself.

Calculate the flux received by the photodetector in the system shown in Fig. 5.11. A lens of 50 mm focal length and NA of 0.5 is used to collimate the light emitted from a 100 μW Lambertian source of 0.1 mm^2 area. The collimated light passes through a medium with an absorption coefficient of 0.75 and is focused onto the detector of 0.1 mm^2 area with a 50 mm lens. Both lenses have a transmission factor of 0.95. [17 μW]

Exercise 5.3

A useful piece of information is that the numerical aperture of a Lambertian source is approximately equal to 1. Verify this for yourself.

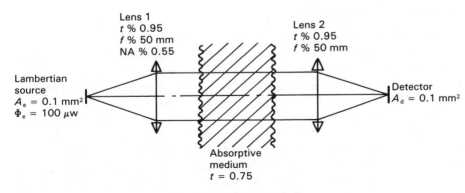

Lens 1
t % 0.95
f % 50 mm
NA % 0.55

Lens 2
t % 0.95
f % 50 mm

Lambertian
source
A_e = 0.1 mm^2
Φ_e = 100 μw

Detector
A_d = 0.1 mm^2

Absorptive
medium
t = 0.75

Fig. 5.11   Exercise 5.3.

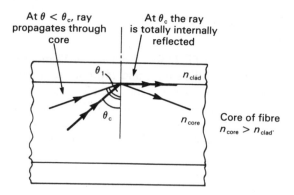

At $\theta < \theta_c$, ray propagates through core

At θ_c the ray is totally internally reflected

n_{clad}

n_{core}

Core of fibre $n_{core} > n_{clad}$.

Fig. 5.12   Critical angle reflection within a fibre.

Fibre Light Guides

The idea of transmitting optical radiation through glass-fibre waveguides was first suggested by two engineers who worked for STL. See the original publication of the work: Kao, K.O. and Hockham, E. *Proceedings of the IEE* Vol. 113, pages 1151–58, (1966).

An alternative means of conveying light from source to detector is via a thin fibre of glass. The concept of light transmission through a fibre treats the fibre as a dielectric waveguide in which the light is guided down the fibre by reflections at the surface of the glass (Fig. 5.12). For guidance of the wave to occur the angle θ at which the ray of light strikes the glass surface must exceed the *critical angle*, θ_c at which *total internal reflection* of the ray will occur. This is expressed as

$$\theta > \theta_c = \sin^{-1}(n_{clad}/n_{core}) \tag{5.25}$$

For a discussion of critical angle and total internal reflection, see Longhurst, R.S. *Geometric and Physical Optics* (Longman, 1974).

where n_{core} and n_{clad} are the refractive indices of the fibre core and the surrounding medium respectively and $n_{core} > n_{clad}$. Total internal reflection occurs at the boundary between two media of different refractive indices when light travelling from the medium of higher refractive index to the lower one strikes the interface at the critical angle. For angles less than this critical angle, the ray is partially transmitted into the second medium and energy is lost. For angles greater than the critical angle the ray is totally reflected back into the original medium. In practice, the core is surrounded by a glass cladding of lower refractive index to protect the core and to prevent changes in the critical angle due to contamination on the surface.

Numerical Aperture of Fibres

We have seen that light which strikes the core/cladding interface at less than the critical angle will not propagate along the fibre. What we really need to understand, however, are the implications that this has on the cone of light which the fibre can accept. In other words, what is the numerical aperture of the fibre?

Refraction is the bending of light which takes place when a ray of light in one medium passes into another medium of a different refractive index.

Light which is launched into the fibre core at an angle θ_i will be refracted at the air/glass interface and will proceed down the fibre at an angle θ_r until it strikes the core/cladding interface at an angle θ (Fig. 5.13). Therefore, if the angle $(90° - \theta_r)$ is exactly equal to the critical angle θ_c, the ray will be *totally internally reflected*. In this case, θ_i will represent the upper limit at which rays can be launched into the fibre. Hence, at the boundary between a medium of refractive index n and another of index n_{core}, the condition of refraction is expressed by

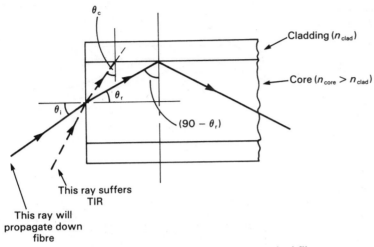

Fig. 5.13 Propagation of rays along optical fibre.

$$n \sin\theta_i = n_{core} \sin\theta_r \qquad (5.26)$$
$$= n_{core} \sin(90° - \theta_c)$$
$$= n_{core} \cos\theta_c$$
$$= n_{core} \sin(1 - \sin^2\theta_c)^{1/2} \qquad (5.27)$$

For critical-angle reflection inside the fibre, from eqn (5.25),

$$\sin\theta_c = n_{clad}/n_{core}$$

and thus,

$$n \sin\theta_i = n_{core}[1 - (n_{clad}/n_{core})^2]^{1/2}$$
$$= (n_{core}^2 - n_{clad}^2)^{1/2}$$

But, we have already seen that

$$NA = n \sin\theta_i$$

therefore, the numerical aperture of the fibre is given by

$$NA = (n_{core}^2 - n_{clad}^2)^{1/2}. \qquad (5.28)$$

Calculate the numerical aperture of a step-index fibre with a core refractive index of 1.53 **Exercise 5.4**
and a cladding refractive index of 1.50.

[0.30]

Types of Fibre

For efficient coupling of light, the active diameter of the source should be less than
the core diameter of the fibre. Even then only some of the light will fall within the
collection cone of the fibre. The greater the numerical aperture, the more light can
be accepted and the better the coupling. In some fibres where there is an abrupt
change in refractive index between core and cladding, *step-index* fibres, the core
diameter is large enough for all rays which strike the core/cladding interface at

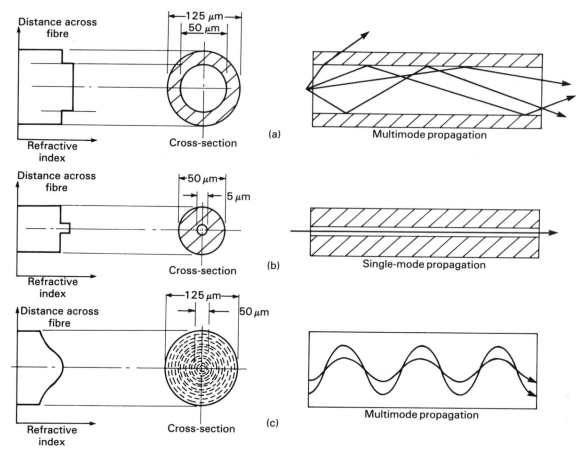

Fig. 5.14 Characteristics of optical fibres. (a) Step-index fibre. (b) Single-mode fibre. (c) Graded-index fibre.

greater than the critical angle to propagate down the fibre independently of each other. In other words, the fibre can support many different modes of propagation: the fibre is known as *multimode*. The overall diameter of multimode fibres is typically 100 to 150 μm with a core around 50 μm diameter (Fig. 5.14). The problem with such fibres is that the different modes travel different distances along the fibre, with the result that they arrive at the end of the fibre at different times. This behaviour is known as *multipath dispersion* and its consequence is that a pulse of light launched at one end will have spread out in time when it reaches the other. The longer the fibre the worse the effect.

These problems can be overcome by restricting the angle of rays which the fibre can accept. This is done by reducing the diameter of the fibre until only one mode can enter and propagate along the fibre (Fig. 5.14). These are known as *single-mode* fibres. The core diameter of a single-mode fibre is the order of the wavelength of light which is to propagate, usually about 2 to 10 μm.

Another solution to the problem of multipath dispersion is obtained by replacing the abrupt refractive-index junction with one which varies gradually from a peak at the centre of the core to its minimum at the interface. These are known as *graded-index* fibres. In this way, rays at the extremities of the acceptance cone travel into

There can be problems with getting enough light into a single mode fibre without damaging the small core. This is particularly true when high-power lasers are used.

regions of steadily decreasing refractive index where they speed up (Fig. 5.14). The net result is that all modes of propagation arrive at the end of the fibre with a much reduced *temporal* spread.

Rays travel faster in regions of low refractive index.

Fibre Bandwidth

The temporal spread of the pulse as it propagates down the fibre has implications as far as the capability of the fibre to transmit information is concerned. The smaller the dispersion, the greater is the potential of the fibre to carry information. For example in high data-rate transmission, a series of pulses of given duration are launched into the fibre at short intervals. For high-dispersion fibres, the pulse smear at the exit end may be so great that the pulses merge together and become totally indistinguishable. There is an obvious limit here to the upper rate at which data can be transmitted.

In the worst case, let us assume that we have one ray travelling down the cladding with the velocity v_{clad} and another travelling axially down the core at a velocity v_{core}. The difference in time for the two rays to reach the other end will be

$$\Delta t = (l/v_{core}) - (l/v_{clad})$$

But since the velocity of a light ray in a medium of refractive index n is c/n, then the time difference is

$$\Delta t = l(n_{core} - n_{clad})/c$$
$$= l\Delta n/c \tag{5.29}$$

where Δn is the refractive-index difference between the two media. If two pulses are launched into the fibre at an interval of less than Δt we would not expect to be able to distinguish the pulses at the other end.

Calculate the maximum pulse spread, due to multipath dispersion, for a 1 km length of step-index fibre outlined in Exercise 5.4.

Worked Example 5.7

Solution: The refractive-index difference between core and cladding for the fibre of Exercise 5.4 is 0.03. Thus for a 1 km length, eqn (5.29) gives the maximum pulse spread to be 100 ns. We would therefore expect that we could not launch pulses into this fibre at faster than 100 ns intervals.

We will return to this point when we discuss fibre-optic communications in Chapter 6.

Attenuation of Fibres

Of particular importance in the application of fibres to optical systems is the amount of light which can be transmitted down the fibre. Several factors contribute to the attenuation of light through fibres, including absorption, scattering and bending effects. Absorption occurs at impurities in the fibre such as beads of water which become imbedded in the glass during manufacture and absorb the light preferentially at certain wavelengths. Scattering is the change in direction of propagation of a ray by collision with an inhomogeneity or irregularity in the glass. This effect is known as *Rayleigh scattering* and is proportional to λ^{-4}. The overall effect is that the total attenuation is wavelength dependent and follows the typical profile shown in Fig. 5.15.

The first optical fibres had attenuations of around 1000 dB km^{-1}. After only 20 years of development, fibres are now available with attenuations as low as 0.1 dB km^{-1}.

Inhomogeneities are local variations in refractive index which alter the speed of the ray and deviate it from its original path.

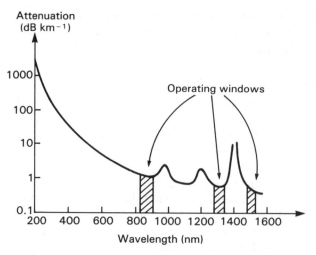

Fig. 5.15 Typical fibre loss characteristics.

Severe bending of the fibre in use can allow light to leak into the cladding and escape when it would normally be reflected. The critical radius at which bending losses become a problem is around two to three millimetres.

Source-to-Fibre Coupling

The simplest method of coupling light into a fibre is merely to butt the fibre as close as possible to the emitting surface of the source (Fig. 5.16). As one might expect this method is likely to result in the greatest loss of light. Provided that the source diameter is less than that of the fibre then the amount of light coupled into the fibre depends solely on the numerical apertures of source and fibre. Thus from eqn (5.20), the coupling efficiency of the fibre is

$$\eta_f = (NA_f)^2/(NA_e)^2 \tag{5.30}$$

Exercise 5.5 Calculate the coupling efficiency when an LED of active area 0.01 mm² is butted against a step-index fibre of 150 μm diameter with a numerical aperture of 0.2. *[Ans 0.04]*

When, as is usually the case, the source's active area is greater than the fibre core then the active areas must be brought into the calculation. The fibre will only intercept that portion of light which corresponds to its core diameter; but only those rays which are within the collection cone of the fibre will propagate. In this case the collection efficiency is given by

$$\eta_f = (A_f/A_e)[(NA_f)^2/(NA_e)^2]t \tag{5.31}$$

We have included a transmission factor t here which represents any reflection losses between source and fibre.

Worked Example 5.8 Calculate the coupling efficiency for an LED with an active diameter of 350 μm

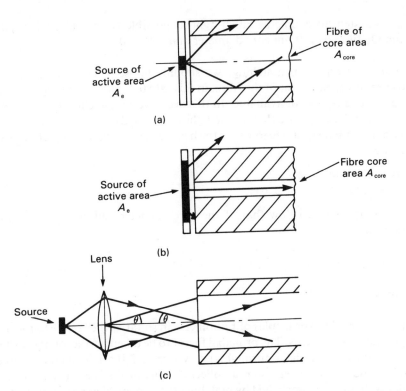

Fibre of core area A_{core}

Source of active area A_e

(a)

Source of active area A_e

Fibre core area A_{core}

(b)

Lens

Source

(c)

Fig. 5.16 Source-to-fibre coupling. (a) Direct (Butt) coupling: $A_e \ll A_{core}$. (b) Direct (Butt) coupling: $A_e \gg A_{core}$. (c) Indirect coupling: $NA_{fibre} = NA_{lens}$.

butted to a step-index fibre with a core diameter of 100 μm and numerical aperture of 0.14. For a glass/air interface the transmission coefficient is about 0.95.

Solution: In the absence of any information about source geometry we must assume that the source profile is Lambertian, hence $NA_e = 1$. The coupling efficiency is given from eqn (5.31) as

$$\mu_m = [(100\ \mu m)^2/(350\ \mu m)^2](0.14)^2(0.95)$$
$$= 1.52 \times 10^{-3}$$

It is usual to express the coupling loss in fibre optics in decibels. Hence, the coupling loss is

$$L_c = 10 \log \eta_f$$
$$= 10 \log(1.52 \times 10^{-3})$$
$$\therefore \quad L_c \sim -28\ \text{dB}$$

The most efficient coupling of light into the fibre will be accomplished by using indirect coupling with a coupling lens. As before, the same conditions and criteria apply as with any other optical component in the system: the numerical aperture of the fibre should be matched to that of the source or coupling lens.

Worked Example 5.9 Design a system to couple as much light as possible from a HeNe laser into a multimode fibre of 100 μm core diameter and NA of 0.2.

Solution: For optimum matching, we should choose a lens with a numerical aperture equivalent to that of the fibre, in this case 0.2. Suppose though that a lens with a suitable NA is not available. In this event we should choose a lens with an NA less than 0.2 to ensure that all the light is collected.

The approximate spot size to which the beam can be focused using a 0.2 NA lens is given by

$$d \simeq 632.8 \text{ nm}/0.2$$
$$\therefore \quad d \simeq 3.2 \ \mu\text{m}$$

This spot size is well below that of the core diameter and efficient coupling is guaranteed.

Summary

We have outlined the principles of radiometry in so far as they relate to the design of optoelectronic systems. The principal factor to be taken into account is to design for optimum flux transfer through the system. Of the two methods of light coupling, indirect transfer using a collection lens offers the greatest collection efficiency. This is accomplished by matching the numerical aperture of the source to that of the detector or collection lens. In addition, for transfer through the atmosphere, fibre light guides may be used. The principles outlined for optimum coupling still apply. Three types of fibre are in common use with different propagation- and refractive-index characteristics, namely:

(i) step index, multimode; (ii) step index, singlemode; (iii) graded index, multimode.

Problems

5.1 A point source of light is placed 1 m from an observation screen of 1 m diameter. If the source intensity is 5 mW sr^{-1}, what is the flux incident on the screen? If a hole of 50 mm diameter is cut in the centre of the screen, what is the fraction of emitted light which passes through the hole?

5.2 The peak flux emitted from a red and a green LED occurs at 650 nm and 560 nm, respectively. Estimate the radiant flux which the red LED would have to emit to produce the same visual response as the green one emitting 10 μW.

5.3 Calculate the flux received by a photodetector of 0.1 mm^2 area, when a lens of 50 mm focal length and NA of 0.5 is used to collect the light emitted from a 100 μW Lambertian source of 0.1 mm^2 area. The lens has a transmission factor of 0.95 and the lens to source distance is 60 mm.

5.4 An LED has an active diameter of 200 μm and an output flux of 100 μW. What is the coupling efficiency, if the source is butt-jointed to a step-index optical fibre of 50 μm core diameter. Estimate also the flux emitted at the exit end of a 100 m length of the fibre. Assume that the fibre has an overall attentuation of 10 dB km^{-1} at the peak emission wavelength of the LED, a core refractive index of 1.53, a cladding index of 1.50 and a diode to fibre transmission factor of 0.95.

5.5 If a small light source which emits a radiant flux of 50 μW is positioned 300 mm from a photodetector with a sensitivity of 0.5 A W^{-1} at the peak emission wavelength of the source, calculate the current generated in the detector. Assume the spatial profile of the source to be Lambertian, the detector area to be 1 cm^2 and the detector is placed normal to the direction of illumination of the source.

5.6 The Lambertian radiation pattern emitted by an LED has half-power points at 65° from its central axis. For a total radiant flux of 10 mW what is the intensity along the central axis? [Hint: solid angle corresponding to half-power points is $2\pi(1 - \cos\theta)$].

5.7 A 150 mm, f/2 lens produces an image of a source of 10 mm diameter at a plane 1 m from the lens. Calculate the size of the image produced. What is the numerical aperture of the lens if its diameter is 50 mm?

5.8 In a two-lens optical system, employing 50 mm diameter lenses with transmission factors of 0.9, the lenses are 50 m apart and have a focal length of 100 mm. The source to lens distance is 100 mm. If the source emits 1 W sr^{-1} from an active area of 1 mm^2, calculate the radiant flux received at the detector plane.

6 Systems and Applications

In this, the final chapter, we bring together all the individual elements and concepts we have outlined so far and discuss the optoelectronic system as a whole. To emphasize the system and design philosophy inherent in these applications we will approach these topics by discussing actual *operational* industrial systems rather than try to give an in-depth treatment of each subject area. The applications are not random choices, however, but have been selected to give as wide a coverage of some of the principal growth area in optoelectronics, namely: fibre communications, fibre sensing, machine vision, laser-scanning systems and holography. Furthermore they have been chosen to display as wide a range as possible of the different types of laser, detector and system configuration available.

The Fibre-Optic Telephone Link

Just as the laser was the catalyst in the growth of optoelectronics, so the development of the optical fibre has proven to be the key element in the rebirth of optical communications. The idea of using light as a means of communication is not new. In pre-historic times the signal fire formed the basis of a crude optical warning system. The eighteenth century saw the development of the signal flag for ship-to-ship communications which, ultimately, gave way to 'Morse' coded signals being flashed out by electric spotlights. The clue to the future of modern *optical communications*, however, came with a demonstration of how speech could be transmitted on a beam of light. Advancement of the idea was impeded for want of a suitable light source and a medium through which light could be transmitted with low loss. Although it is possible to communicate optically through the atmosphere, attenuation by dust and water vapour severely limits the path length of the communications link. Furthermore, since a direct line of sight is generally needed between transmitter and receiver, its potential application is limited.

By Alexander Graham Bell, the inventor of the telephone, in 1880. His original idea has been revived and forms the basis of a sensitive spectroscopic technique known as photo-acoustic spectroscopy.

Kao, K.O. and Hockham, E. in *Proceedings of the IEEE* Vol. 113, pages 1151–8, (1966).

The need for a medium in which light could be guided at low loss was apparent. The impetus for a serious re-appraisal of optical communications came in 1966 with the suggestion of using a length of glass fibre as an optical waveguide. Initially the light loss in such fibres was as high as 1000 dB/km. An intensive period of development brought this figure down to around 10 dB/km by the mid-1970's. With this outstanding progress in the field of materials science, coupled with the parallel improvements made to the lifetime and reliability of semiconductor lasers and LEDs, the time was ripe for optical communications to become a practical reality.

To illustrate the features and development of optical communications in the UK we will now look at a typical application: the optical telephone link. Before discussing the system elements themselves it would be useful, first of all, to discuss why using light presents any benefit over conventional communications systems. In other words we need to ask similar questions to those posed in Chapter 1 when we discussed the CD system, namely, *why use light to transmit information?*

Basic Communications Principles

The underlying principle in all telecommunication systems is the desire to transmit what may be basically low-frequency disturbances, such as sound, over large distances. Sound, though, is limited as the basis of a communications network since it will not carry over distances greater than a few kilometres, and even then only in favourable conditions. Also the relatively low velocity at which it travels is a hindrance to rapid communication. In all modern communications system sound is converted into an electrical disturbance using some form of transducer with the immediate advantage that the signal now travels as an electromagnetic wave. Accordingly, transmission distances by this means can be several hundred kilometres.

To enable several pieces of information to be transmitted *simultaneously*, the signal to be transmitted, the *message*, is superimposed in some way onto a reference signal, the *carrier wave*. The carrier wave is *modulated* by the signal wave. In this way, messages which are composed of primarily low frequencies are encoded onto reference signals of high frequency. The transmission of the message is therefore governed by the properties of the carrier and not by those of the message itself. For example, speech and music signals, which are concentrated in the range 50 Hz to 20 kHz, can be transmitted through the atmosphere by a radio wave of several-hundred kiloHertz or higher. The higher the frequency of the carrier signal, the more messages can be encoded onto it. The original messages are ultimately recovered by suitable processing of the signal.

Several forms of modulation, based on both analogue and digital methods, are employed in communications systems. The simplest method of analogue modulation, *amplitude modulation (AM)*, is widely used for the transmission of speech and other low-frequency signals. In its basic form the amplitude of a high-frequency sinusoidal carrier wave is modulated by the message (Fig. 6.1). We can now see why light is so useful as a medium for the transmission of information. Because of its inherently high frequency, around 100 THz for visible light, there is almost a million times more capacity to encode messages onto it than there is for radio frequencies of a few megaHertz. A worked example will illustrate this.

Good overviews of optical-fibre communications can be found in Giallorenzi, T.G. in *Proceedings of the IEEE*, July 1978. For more comprehensive treatments see Senior, J.M. *Optical Fibre Communications* (Prentice-Hall, 1985) and Gowar, J. *Optical Communication Systems* (Prentice-Hall, 1984).

Television and data are other forms of information which may be transmitted by communications links.

The velocity of sound at sea level is about 330 m s⁻¹.

For a discussion of the fundamentals of communications see O'Reilly, J.J. *Telecommunications Principles* (Van Nostrand Reinhold, 1984).

How many telephone channels could, theoretically, be transmitted via an optical communications link using simple amplitude modulation?

Solution: In telephone systems, the frequency range required to transmit recognizable voice signals is from a few hundred Hertz up to 4 kHz. Although voice signals cover the whole audio range up to 18 kHz, the spectrum is composed primarily of low frequencies. From the *Nyquist criterion* discussed earlier the

Worked Example 6.1

See Chapter 1.

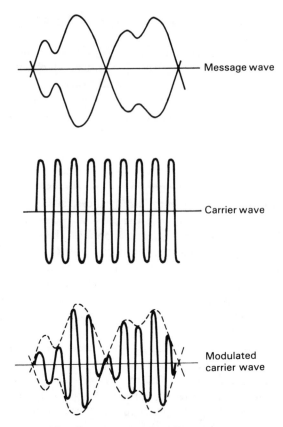

Fig. 6.1   Amplitude modulation.

system should be designed to accommodate a bandwidth at least twice that of the highest frequency to be transmitted. Hence, for voice messages a minimum bandwidth of 8 kHz is needed. For simplicity we can assume 10 kHz. Typically, light in the visible part of the electromagnetic spectrum has a frequency of about 100 THz. Thus, the number of channels which can be modulated onto a light beam is, approximately,

$$(100 \times 10^{12}\,\text{Hz})/(10 \times 10^3\,\text{Hz}) = 10^{10}$$

This is an extremely simplified calculation and completely disregards such factors as fibre bandwidth and detector response. Compared with an all-electrical system, though, in which the number of channels is limited by the use of comparatively lower radio frequencies as the carrier signal, such a simplified calculation does show the massive potential of the optical system.

In addition to the high information capacity, use of light rather than an electronic signal reduces the susceptibility to radiated interference, both externally and between channels.

System Elements

The basic arrangement of an optical-fibre telephone link (Figure 6.2) is similar to

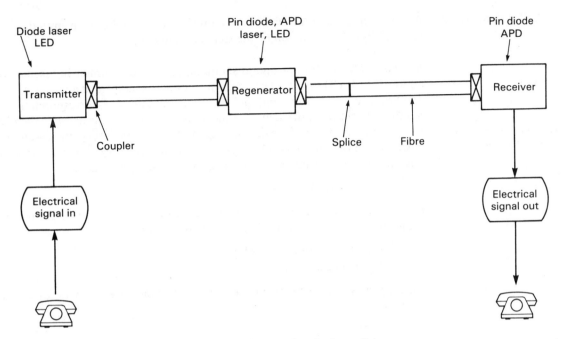

Fig. 6.2   Fibre-optic telephone link.

the coaxial cable systems which have been the standard for telephone links in the UK for many years. The principal features of the optical system are that the *transmitter* terminal consists of a light source and modulator with associated power supplies and drive circuitry. The optical fibre carries the encoded signal to the *receiver* terminal, which consists of photodetector and demodulator again with associated power and drive circuitry. Depending on the type of fibre used and the overall length of the route, successive levels of regeneration of the transmitted signal may be required. Each regenerator consists of an optical receiver directly connected to a transmitter without the need for a modulator and demodulator. One of the most significant differences between optical and coaxial systems is the considerably enhanced spacing between regenerators.

See articles by Martin-Royle, R.D. and Bennett, G.H. in *British Telecommunications Engineering* Vol. 1, pages 190–9, 1983 and by Muir, A.W. in *British Telecommunications Engineering* Vol. 2, pages 180–5, 1983 for excellent overviews of optical-fibre systems in the British Telecom network.

Digital Modulation

In recent years, digital modulation, sometimes known as *pulse-code modulation (PCM)*, of the carrier wave has come into strong contention as the principal method of signal encoding in telephone links, in which the amplitude of the message is encoded as a series of binary pulses. The prime advantage of digital modulation over analogue modulation is the freedom from accumulation of noise and distortion. A signal consisting of a series of equal-amplitude pulses even when badly distorted by noise is easy to recover from the original signal. Again, as in the design of CD systems, we find that the application of optics and the use of digital-signal processing go hand-in-hand. The designers of optoelectronic telephone links decided that the unique advantages offered by the use of light as the carrier medium were best exploited if digital modulation of the signal was used. The capabilities of a digitally modulated optical link are shown by way of an example.

See Beckley, D.J. Main Transmission Network: Planning of Digital Transmission Systems, *Post Office Electrical Engineering Journal* Vol. 72, pages 27–35, 1979.

Worked Example 6.2   Estimate the number of telephone channels which could be transmitted on a light beam if digital modulation of the signal is employed.

See Fig. 1.6.

8-bit quantization is the norm in telephone communications.

Solution: The method of digital sampling employed is similar in essence to that adopted in the CD system. The analogue message is sampled at a fixed rate and the amplitude at each sampling point converted into the corresponding binary number. Again from the sampling theorem, we must sample at twice the highest-frequency component of the message signal. Hence, converting into a 8-bit number means we must sample at 16 times the highest frequency. For telephone transmission, the upper frequency is 4 kHz, thus the sampling rate is

$$4 \times 10^3 \, \text{Hz} \times 16 = 64 \, \text{kbit s}^{-1}$$

For an optical system based on a carrier frequency of 100 THz, the number of channels which can be accommodated is, therefore,

$$100 \times 10^{12} \, \text{bit s}^{-1}/64 \times 10^3 \, \text{bit s}^{-1}$$
$$\sim 10^9 \, \text{channels of data.}$$

Once again we have taken no account of system bandwidth or of any other limitations.

The above figure gives the approximate theoretical limit to the number of simultaneous data channels which can be transmitted. In practice, this capacity is still out of reach. In the UK, a number of optical-fibre routes have been installed based on internationally designated transmission bit rates of 8, 34, 140 and 565 Mbit/s.

Light Source

The requirements which have to be met for a light source to be suitable for use in optical-fibre communications are manifold. Firstly, it should have a high irradiance to compensate for transmission losses in, and coupling losses into, the fibre. Transmission losses will be minimized if the peak emission wavelength of the source matches the optimum transmission window of the optical fibre. Secondly, the source should exhibit low beam divergence and high spectral purity in order that the beam can be focused to a small spot size, once again minimizing coupling losses. Furthermore, high spectral purity reduces the spread of the light pulse as it propagates down the fibre. Finally, to facilitate easy use and installation, the device should be physically small.

See Chapter 5 for a discussion of transmission and coupling losses.

Remember our discussion of dispersion in Chapter 5.

Although radiating generally less flux than lasers and having a wider spectral linewidth, LEDs meet many of the requirements listed above and are cheaper and easier to utilize. Drive circuitry is less sophisticated and optical design is simpler. Edge emitting LEDs, for example, have the benefit of higher output powers without the need for the close current control required by lasers. Lasers are best suited to higher-capacity systems and are essential when monomode fibres are used.

Additionally semiconductor sources require only simple driving circuits and can be directly modulated through their drive current.

The Optical Fibre

The over-riding advantage to be gained from the use of fibres is, simply, the ability to guide the signal, with low loss, anywhere we want rather than be restricted to line-of-sight communications. With this advantage in mind we can contemplate the replacement of existing telephone links with their fibre equivalent: we could even use existing ducting. Since fibres are typically 125 μm diameter, overall cable dimensions are substantially narrower than standard coaxial cabling even when allowance is made for protective sheathing. These small diameters ease the installation and simplifies storage.

The question remains as to which type of fibre should be used, step-index, graded-index or monomode? The simple step-index fibre is relatively easy to couple light into but suffers badly from multipath dispersion which ultimately limits the system bandwidth, and is not generally used in long-distance telecommunications.

Worked Example 6.3

We can estimate simply what the upper bandwidth might be if multipath dispersion were the only mode of pulse spreading in the fibre.

Solution: In the worst case let us assume we have one ray travelling down the cladding at velocity v_{clad} and another ray travelling axially down the core with a velocity v_{core}. The difference in time for the two rays to reach the end of the fibre will be, from eqn (5.29),

$$\Delta t = l\Delta n/c$$

where l is the length of the fibre and Δn is the difference in refractive index between core and cladding. For a fibre of 1 km length and Δn of 0.03 the pulse spread is about 100 ns. This would give an upper limit to the data transmission rate of about 10 Mbit s^{-1}.

Graded-index fibres present much better control over pulse spreading and are therefore favoured for all low-to-medium capacity links. Minimum dispersion is obtained with monomode fibres with accordingly wider bandwidth and are therefore used for high-capacity systems. Dispersion is also related to spectral purity and high-capacity systems must be used in conjunction with lasers. With modern low-loss fibres, large repeater spacings are possible.

The Detector

Two principal types of detector are commonly found in optical communication systems, namely, avalanche photodiodes (APDs) and pin photodiodes, the main features of which have been outlined earlier. In essence, APDs operate with high reverse potentials of around 300 to 400 V. Furthermore the breakdown voltage increases with temperature which ultimately leads to a decrease in the quantum efficiency of the device. Against this, pin diodes need only reverse potentials of less than 20 V. The sensitivity of the basic pin diode, which is where it suffers in comparison with the APD, can be improved by coupling a pin chip and a low-noise, low-capacitance FET amplifier chip on the same substrate. These hybrid devices, PINFETs, are now the accepted form of pin diode for long-haul, medium-to-high data-rate routes, particularly at wavelengths of 1300 nm.

See Chapter 4.

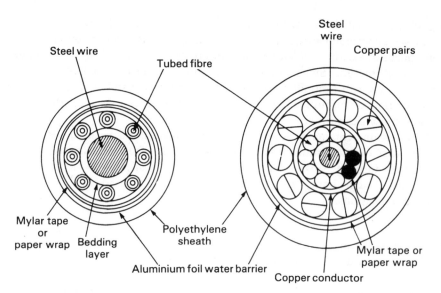

Fig. 6.3 Fibre-cable construction (reproduced by permission of British Telecom).

Proprietary Optical Links

The historical development of fibre-based telephone links in the UK is interesting to follow. After some early experimental routes in 1977, a number of proprietary links were installed by the Post Office Telecommunications, now British Telecom, primarily at bit rates of 8, 34 and 140 Mbits s^{-1}. The routes were selected to develop and advance the optical fibre technology in the UK, and to provide a variety of conditions under which systems must operate. All the initial systems used graded-index fibres. Each cable, shown in Fig. 6.3, contained 8 fibres.

The first operational link was Brownhills to Walsall, installed in autumn 1980. A later link, brought into operation in April 1982, was that between Aberdeen and Kingswells. This route is typical of the early 8 Mbit s^{-1} systems and consisted of four links 12.3 km long with no intermediate regeneration.

The Birmingham to London route is representative of the higher-capacity 140 Mbit s^{-1} systems. This route was upgraded in 1984 from an existing 34 Mbit system. The total length of each link is 205 km with regeneration provided at eight surface and sixteen buried stations. The maximum section length between regeneration is 10.3 km.

Also around this period, the first link to use single-mode fibres was installed: a 140 Mbit s^{-1} link between Luton and Milton Keynes. The distance of 27 km is covered without regeneration. More recently, the first operational 565 Mbit s^{-1} link was installed towards the end of 1985 between Nottingham and Sheffield. The overall length is 72 km with the longest section between regeneration being 26 km.

A summary of these routes is given in Table 6.1. These development links, and others, gave manufacturers the opportunity to assess the economic viability of optical-fibre links in the context of the UK telephone network.

A Fibre-Optic Current Sensor

Initially an offshot of the fibre-communications programme, the development of

Table 6.1
Characteristics of typical optical-fibre links

	Aberdeen-Kingswells	Birmingham-London	Nottingham-Sheffield
Bit rate	8 Mbit s^{-1}	140 Mbit s^{-1}	565 Mbit s^{-1}
Operational wavelength	850 $\pm$ 10 nm	1285 $\pm$ 15 nm	1300 nm
Transmitter	Laser diode (STC TS 8040)	LED	Heterostructure laser diode
Receiver	APD (RCA C30021)	PINFET	PINFET
Fibre	Graded index	Graded index	Single mode
No. of fibres in cable	8	8	8
Fibre dimensions	50/125 μm	50/125 μm	8/125 μm
Attenuation	3dB km^{-1}	1–1.5 dB km^{-1}	0.5dB km^{-1}
NA	0.2		
System length	12.3 km	205 km	72 km
Repeater spacing	None	~10 km	26 km
Launch power	−6 dBm	−9 dBm	−3 dBm
Diode sensitivity	−62 dBm	−43 dBm	−36 dBm
Operational date	Spring 1982	Summer 1984	Autumn 1985

fibre-optic sensors is one of the most exciting areas of current optoelectronics research. The basic theme of the fibre sensor is that light passing through an optical fibre is acted on, in some way, by external conditions with a consequential change in the properties of the transmitted light. In some instances, the fibre itself may not play a part in the sensing mechanism but merely guides the light to the region of interest.

Figure 6.4 shows a simple fibre sensor which will detect liquid levels in a vessel. In the absence of any liquid in the vessel, light is returned from the end of the fibre because of total internal reflection at the wedge-shaped tip. As the liquid rises to cover the end of the fibre, there will be a change in the amount of light reflected. Some will, in fact, leak into the liquid because of the decreased refractive-index differential between fibre and the surrounding medium. Consequently, there is a drop in the amount of light arriving back at the detector.

Some other examples of fibre sensors are known diagrammatically in Fig. 6.5.

> The phenomenon of total internal reflection was discussed in Chapter 5.
>
> Some excellent reviews of fibre sensing are to be found in Pitt, G.D. *et al, IEEE Proceedings — J, Optoelectronics*, Vol. 132, pages 214–48, 1985; Rogers, A.J., *CEGB Research*, pages 3–17, 1984; Dakin, J.P., 'Optical Fibre Sensors' in *Proceedings of Electro-optics/Laser International Conference* (IPC 1982). There is also an excellent book by Culshaw, B., *Optical Fibre Sensing and Signal Processing* (*IEEE*).

By examining Fig. 6.5, discuss the operation of the various types of fibre sensor shown.   **Exercise 6.1**

The advantages to be gained by using fibre sensors for such applications parallel, in many ways, those for optical communications. Fibre sensors are generally less prone to damage at high temperatures than conventional measurement devices, do not suffer from electrical interference and, since they do not carry electric current, are *intrinsically safe*. All in all, the fibre sensor is ideally suited for use in hazardous environments or in areas where there may be explosive risks, such as in the offshore oil and gas industries or the electricity supply industry. Additionally, because optical measurement techniques generally make no physical contact with the

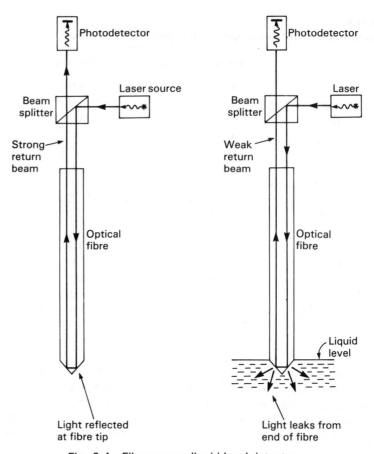

Fig. 6.4 Fibre-sensor liquid-level detector.

measurement area they can be operated remotely and non-destructively. A non-exhaustive list of physical parameters which have been monitored using fibre sensors include: temperature, pressure, strain, liquid level, acceleration, rotation, current, voltage and electromagnetic fields.

Current Sensing

The measurement of current flowing in electrical conductors is one area where the application of fibre sensors can bring significant advantages. This is particularly true when the currents are the order of a few-hundred amperes, such as in electrical power transmission. Conventional methods of current measurement in use by the electricity supply industry employ bulky transformers which are difficult to insulate and have a fairly slow response to transients. Optical techniques, by contrast, can be compact, inexpensive and have a fast response.

The Central Electricity Generating Board have for some years now been investigating the application of fibre sensors to the monitoring of current and voltage in power-transmission systems. Here we will discuss the operation of a prototype current sensor shown diagrammatically in Fig. 6.6.

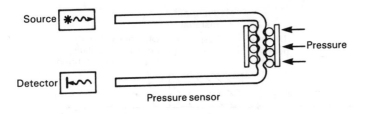

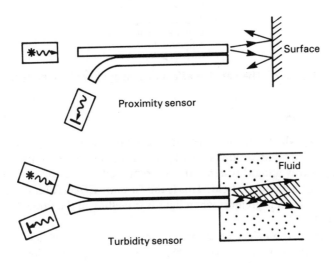

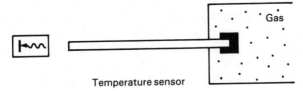

Fig. 6.5   Some typical fibre sensors.

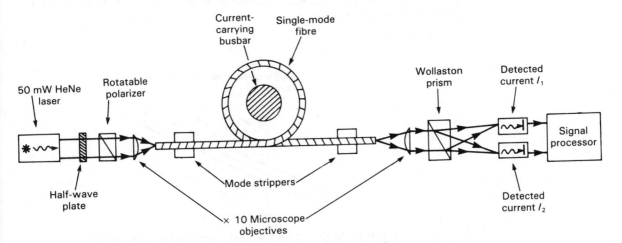

Fig. 6.6   Fibre-optic current sensor (adapted from an original diagram published by the CEGB).

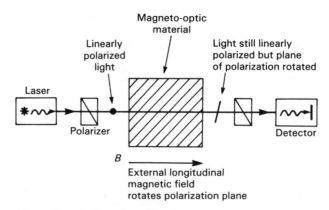

Fig. 6.7　The Faraday effect in a magneto-optic material.

Sensing Principles

The method by which current can be measured by an optical fibre is based on the principle that the polarization (see Chapter 2 for a discussion of polarization) of a beam of light travelling through a transparent medium can be acted upon by the introduction of a magnetic field. This is the *Faraday Effect* (Fig. 6.7). The magnitude of the rotation is greatest when light is travelling in the same direction as the magnetic field.

The Faraday effect was one of the earliest indications of a connection between electromagnetic fields and light.

The origin of the effect arises from the way a magnetic field influences the atomic absorption of a material. This leads to a slightly different velocity of propagation for circularly polarized light depending on whether it is left or right polarized. Since linearly polarized light can be considered as equivalent to a combination of two circularly polarized components of opposite sense, we can see that if one component propagates faster than the other as the beam propagates through the medium there will be a net rotation of the polarized beam.

See the classic work: Born, M. and Wolf, E. *Principles of Optics* (Pergamon, 1980).

Turning our attention back now to our knowledge of basic electricity and magnetism, we know that a current-carrying conductor will give rise to a magnetic field acting at right angles to the current direction (Fig. 6.8). If an optical fibre, carrying a linearly polarized light beam, is wrapped around the conductor the action of the magnetic field will rotate the plane of polarization of the beam. The angle of rotation is given by

$$\beta = k_{\mathrm{v}} H x \tag{6.1}$$

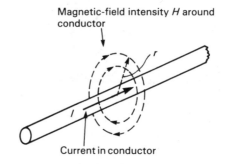

Fig. 6.8　Magnetic field around current-carrying conductor.

where k_v is a constant, known as the *Verdet* constant for a given material, H is the magnitude of the magnetic field and x is the path length travelled by the beam in the magnetic field. Basic theory tells us that the magnetic-field intensity associated with a current I flowing in a conductor is given by

Refer to Compton, A.J. *Basic Electromagnetism* (Van Nostrand Reinhold, 1986).

$$H = I/2\pi r \qquad (6.2)$$

where r is the distance between the conductor and the point of measurement of the field intensity. Hence, the Faraday rotation can be written as,

$$\beta = k_v(I/2\pi r)x$$

But x, the path travelled by the light, will also be equal to $2\pi r$ if the fibre is looped once around the conductor, and therefore,

$$\beta = k_v I \qquad (6.3)$$

Provided that the fibre is wound around the conductor at a uniform distance, eqn (6.3) gives the magnitude of the rotation. Having several loops of fibre around the conductor increases the path length of the beam and enhances the magnitude of the effect and compensates for stray magnetic fields.

Experimentally, the Verdet constant for this system is found to be 15.6×10^{-3} min A^{-1}. Hence, the Faraday rotation for a current of 1000 A r.m.s. will be 15.6 minutes of arc.

The Sensing Element

Selection of the fibre is crucial to the operation. Since it is a measurement of polarization rotation that we wish to make, any natural change of the polarization state of the beam as it propagates along the fibre could mask the effect we are trying to measure. Such materials are said to be *bi-refringent* and all fibres exhibit some degree of intrinsic bi-refringence. There is an optimum operating point for the fibre, however, corresponding to a preferred plane of polarization which the incident beam should possess such that the exit beam will emerge linearly polarized, but not necessarily in the same plane. This condition is achieved by placing a *half-wave plate* in front of the fibre which will allow the polarization plane of the incident light to be rotated into the preferred polarization plane for the fibre. Specially manufactured *low-bi-refringence* fibres are available which minimize any intrinsic effects.

For further expansion of this point see Rogers A.J. in *IEE Proceedings* Vol. 132, pages 303–8, 1985.

A half-wave plate is a piece of bi-refringent material, such as quartz, precisely manufactured to a thickness of exactly one-half of the wavelength of light being used. Rotating such a plate also rotates the plane of polarization of the incident light.

Furthermore, since the propagation along the fibre of more than one mode will tend to complicate the polarization pattern, magnetic rotation will be averaged out. This eventuality dictates the choice of single-mode fibres. Even so, unwanted modes will still manage to propagate along the fibre due to the launching of some available light into the cladding. These unwanted modes can be removed by passing the fibre through a short length of liquid whose refractive index is higher than that of the fibre such that the unwanted modes are coupled out.

In case other unwanted modes are generated in the fibre due to bending or pressure further mode stripping can be carried out at the exit end of the fibre.

Suitable liquids are paraffin, $n = 1.48$, or turpentine, $n = 1.47$. Compare with the fibre refractive index of 1.458.

Launching Optics

The selection of a monomode fibre creates problems with launching enough light into it to yield a significant signal at the other end. A HeNe laser delivering a linearly polarized output of 50 mW continuous power is passed through a polarizing prism to enhance the degree of polarization to about 1 part in 10^4. The beam is launched into the fibre using a microscope objective lens of 10x magnification and numerical aperture of 0.12. Comparing the numerical aperture of the lens with that of the fibre would appear to indicate a mismatch between the two resulting in only a fraction, $0.048/0.12 = 0.40$, of light being coupled into the fibre. The actual situation, however, is not as bad as it appears at first sight; since the laser-beam diameter is only about 2 mm and the entrance diameter of the lens 8 mm, only one-quarter of the useful diameter of the lens is utilized. The effective NA is therefore about 0.03 and efficient launching of the beam is possible.

Exercise 6.2 Calculate the optical power received at the end of 6 m of single-mode optical fibre when light from a 50 mW HeNe laser is launched into it using the optical system outlined above. The core diameter is 7 μm, its numerical aperture 0.05 and its attentuation 50 dB km^{-1}.

Detection Optics

To measure rotation we need to compare the relative orientation of polarization between the output and input beams. This is accomplished by collecting the emitted light by another 10x/0.12 NA objective and passing it into a *Wollaston prism*. Such a prism resolves the beam into two separate linearly polarized components at right angles to each other. Each component is detected by separate pin photodetectors, the outputs of which are connected to sum-and-difference amplifiers to produce an output proportional to

$$(I_1 - I_2)/(I_1 + I_2)$$

where I_1 and I_2 are the respective photocurrents produced in the diodes. Such a procedure is the classical method for ensuring maximum output sensitivity and also ensuring that the output is independent of beam-intensity fluctuations and detector differences. The prism may also be rotated to gain the optimum output signal from the detectors.

In the prototype instrument, a linear response of detector output against busbar current was obtained for the range zero and 1000 A r.m.s.

Optical Imaging Using CCD Cameras

See Chapter 4.

The detectors we discussed earlier have one characteristic in common, they integrate in space and time the total amount of light falling on them. For example, if a beam of light of given radiant flux falls on a small area of a photodiode surface it will produce the same response as a beam of the same flux covering the entire diode area (Fig. 6.9). Overfilling the detector with light, however, will produce a response proportional to the radiant flux per unit area of the detector. Moving the detector across the light field produces a response which varies with the point-by-point irradiance of the incident light. In many applications of optoelectronics it is essential to preserve any spatial character that the light wave may have. Detectors which accomplish this are known as *imaging detectors*.

The ubiquitous photographic emulsion is probably the best known, most fre-

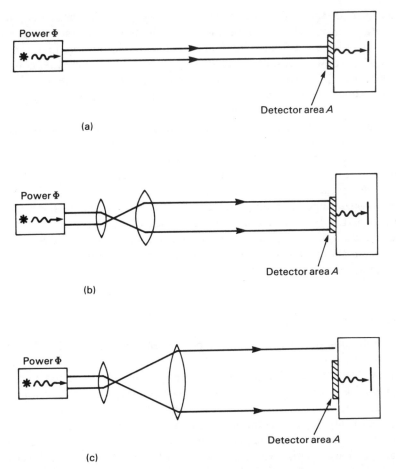

Fig. 6.9 Light coupling into detector. (a) Detector underfilled: output proportional to Φ. (b) Detector just filled: output proportional to Φ. (c) Detector overfilled: output proportional to Φ/A.

quently used and undervalued of imaging detectors. Each clump of silver-halide grains, of which the emulsion is composed, will respond separately to the irradiance of light falling on it. After chemical processing, the emulsion retains a permanent image of the incident light. It is this archival permanence of photographic film which is its most important characteristic. Others are its high resolving capabilities over a wide spectral range coupled with low cost. Against this must be weighed its low dynamic range, non-linearity, lack of real-time response and the requirement for wet processing.

The *vidicon tube* is another example of a widely used imaging detector. Based on the principles of the cathode-ray tube, an illuminated photoconductive target is scanned by an electron beam to produce a variation in the electric current drawn by it. In its many derivative forms, the vidicon offers real-time imaging over a wide spectral range and with a moderately high quantum efficiency. Its disadvantages include a low signal-to-noise ratio, susceptibility to retention of some of the signal after readout (known as *lag*) and leakage of the signal from intensely illuminated areas into neighbouring areas (known as *blooming*). Cameras based on the vidicon and its derivatives are standard for television broadcasting.

For a more detailed discussion of the photographic emulsion see Harder, A. (ed.) (1971) *The Manual of Photography* (Focal Press) or Mees, C.E.K. *The Theory of the Photographic Process* (Macmillan).

Other commercial forms are the *newvicon, plumbicon* and *orthicon*.

For further discussion of TV cameras, see Bohlman, K.J. *Closed Circuit Television for Technicians* (N. Price, 1978).

The future of real-time image detection, however, lies with the current breed of *solid-state imagers* which are beginning to dominate the industrial market. Cameras based around solid-state imagers offer high resolution, high sensitivity, small size, low power consumption and easy interfacing for computer control and data handling. Such cameras are rapidly gaining favour for image processing, machine vision and industrial surveillance and control. One approach to solid-state imaging is to fabricate many tiny photodiodes together in a linear or two-dimensional array. Fabrication difficulties, however, and low packing densities have led to the development of the *charge-coupled device* (CCD). As the name implies, the CCD carries its information by way of efficient transfer of packets of electric charge through the device. The advantages offered by CCD technology are the increase in packing density afforded over that of the photodiode array, the freedom from lag and blooming and the sensitivity at low light levels. Perhaps most significant is the freedom from geometrical distortion of the image which is a severe limitation of vidicon cameras. It is on this device that much of the future of solid-state imaging depends and, therefore, our discussion will concentrate on CCD cameras and their application.

Charge Storage

n-type is equally suitable but applied voltages will be negative.

See Chapters 3 and 4 for discussions of charge carriers.

Consider the situation where we have a p-type semiconductor with a thin layer of silicon dioxide deposited on it. A metal electrode is deposited on the oxide layer. Before application of a voltage, holes will be evenly distributed across the material. On applying a positive voltage, holes are pushed away from the area under the electrode: a depletion layer is formed. Increasing the voltage increases the extent of the depletion layer. At a certain voltage the surface potential is so high that electrons are attracted to the surface, whereupon, they form a thin but dense layer of negative charge. This creates an *n-channel* CCD. The depletion layer can be

A *p-channel* CCD is formed from an n-type semiconductor.

thought of in terms of a *potential well*, a region of low potential in the material below the plate, acting as a sink for electrons (Fig. 6.10a). The oxide layer between the electrode and the silicon helps prevent drift of charge away from the well. The device as described here is a *surface-channel* CCD. All practical devices use a *buried-channel* structure in which charge is stored well below the surface for more efficient transfer. The concepts, however, are similar.

Charge Transfer

The charge packet contained in the well can be moved across the device by successive applications of a voltage pulse to the interconnecting electrodes (Fig. 6.10b–e). Initially we assume that charge is contained under the first electrode. When a potential is applied to the second electrode the width of the well increases across the region under both electrodes. The charge now becomes shared between the two electrodes. Removing the voltage from the first electrode causes that well to collapse, pushing all the available charge into the second well. The well and its associated charge have moved through the semiconductor. It is worth pointing out the accumulated charge can be stored for several hours without appreciable decay. The efficiency of the charge transfer process approaches a remarkable 99.999%!

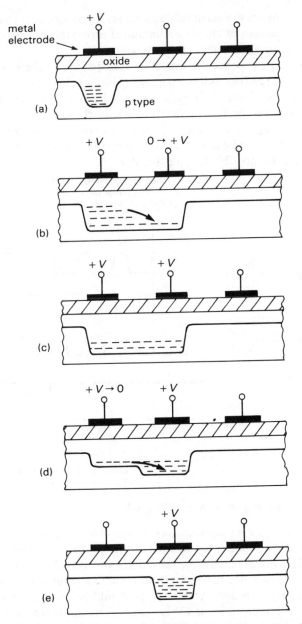

Fig. 6.10 Movement of charge packet between electrodes. (a) Potential well containing charge. (b) Adjacent well created: charge 'spills' over. (c) Charge shared between wells. (d) First well collapsing. (e) Charge packet has moved between electrodes.

Optical Imaging

As we have already seen many semiconducting materials such as silicon are photo-sensitive. Incident light generates electron-hole pairs in the material. The electrons are trapped in the potential well whereas the holes disappear into the substrate. If we consider an n-channel CCD, the number of electrons produced over a given

123

period of time, and hence the accumulated charge, is proportional to the incident light intensity. The pattern of charge accumulated across the device replicates the variation in light intensity of the original image.

Readout of the signal is accomplished by transferring packets of charge to adjacents CCDs to be read out as a video image. Although in principle, each sensor could be fed from a separate clock pulse, in practice the electrodes are grouped together in sets of three or four, called *phases*. Each phase is connected to a separate clocked voltage. For the system outlined here a *three-phase* clocking system allows the charge to be moved entirely across the device (Fig. 6.11). Clock rates up to 10 MHz are possible for line transfer.

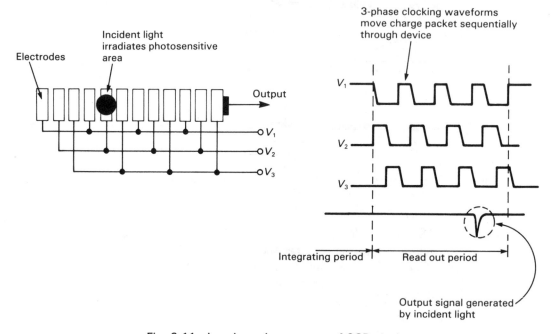

Fig. 6.11 Imaging using an array of CCD pixels.

The simple line-imaging arrangement as shown previously is not suitable for area detectors because of the time taken to transfer the charge across the device. For area sensors, the individual pixels are arranged in a matrix of $M \times N$ elements. The packets of charge are switched to a series of stores and readout sequentially while a second image is being recorded. Area imagers should be able to transfer data at rates compatible with conventional TV systems. There are two principal ways in which information is organized on a CCD chip, namely, *frame transfer* and *inter-line transfer*.

In frame transfer the array is divided into vertical columns with channel stops in between (Fig. 6.12). Electrodes run across the array at right angles to the columns. The array is divided into two sections, a storage region which is shielded from light and an optically sensitive region. The basic principle of operation is that the entire sensitive area is exposed to light and then the developed charge is transferred to the storage area for readout. This is necessary for the fast readout rates needed in TV systems and only then because of the time taken for readout, not for data transfer. The data is read out sequentially while the next frame is being recorded.

In interline transfer, readout sections are interspersed with sensitive elements.

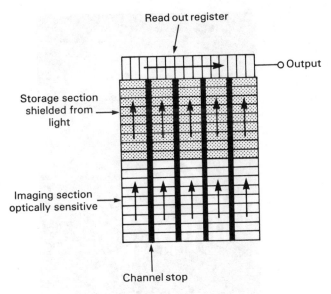

Fig. 6.12 Frame transfer CCD.

The output from each column is taken in parallel using a common register (Fig. 6.13). Although potentially faster than frame transfer, because the light sensitive regions are interleaved with non-sensitive regions, half the image points are lost.

Machine Vision

A typical automated system for machine vision using a CCD camera is shown schematically in Fig. 6.14. A computer-controlled frame store, often called a *frame grabber*, is at the heart of operations. It is here that the electrical signal from

Although we are concerned here with the use of CCDs in TV systems, because of their long storage times, they can be used in a 'slow-integrating' mode which is invaluable in astronomy. It was this application which provided the early impetus for CCD development.

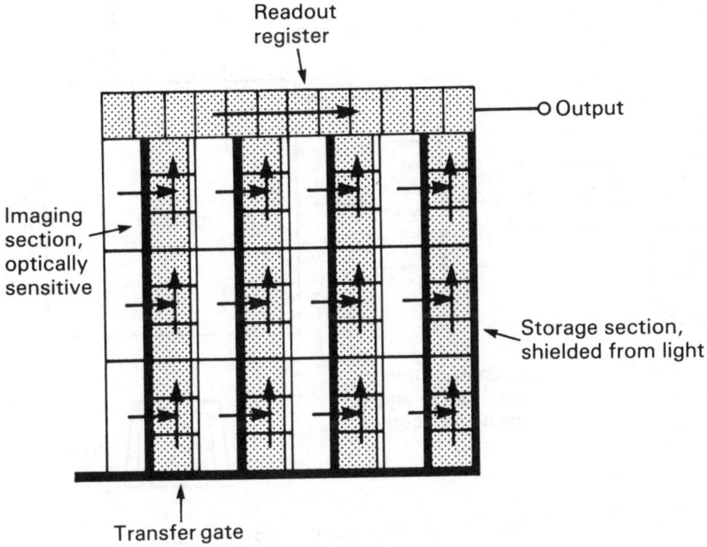

Fig. 6.13 Interline transfer CCD.

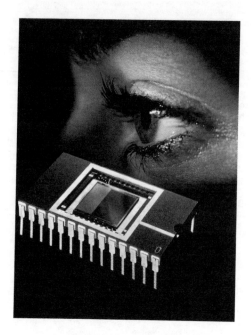

Fig. 6.13 continued

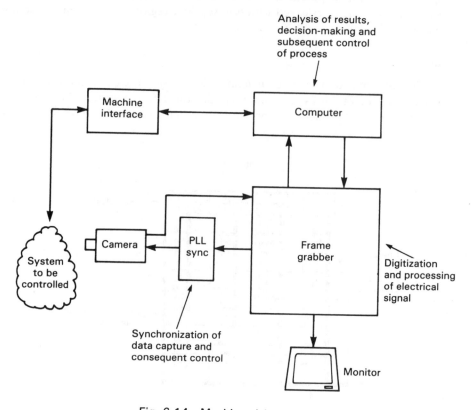

Fig. 6.14 Machine vision system.

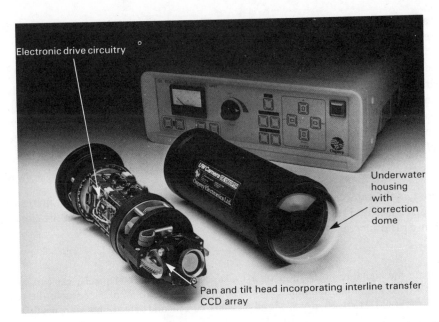

Fig. 6.15   Industrial CCD camera customized for underwater use. (Reproduced by permission of Osprey Electronics Ltd.)

the CCD camera is digitized and processed before being committed to the computer for analysis and subsequent decision making. Synchronization pulses ensure that the camera operates in phase with any processing and control system. Although cameras based on CCD detectors do not yet meet the exacting standards of broadcast TV in the UK, the resolution requirements for industrial machine vision systems are less demanding.

The industrial standard for television, known as CCIR, is based on a format of 576 horizontally-scanned lines composed of two interlaced fields of 288 lines. Current state-of-the-art CCDs, for example the Philips NXA 1010, contain up to 294 lines of 604 pixels each. Allowing a few lines for field blanking this leaves effectively a total of 347 904 pixels, each of 10 μm $\times$ 15.6 μm, for imaging. A typical camera is shown in Fig. 6.15.

Signal-to-Noise Ratio

The ideal image sensor for machine vision should have high sensitivity and a wide dynamic range. As in all detectors the sensitivity of the device is related to the presence of noise such as dark current and shot noise. We saw earlier, that the rate of photoelectron production at a photosensitive detector is given by

Eqns (4.3) and (4.4) in Chapter 4.

$$n_e = \eta \Phi \lambda / hc$$

Since the sensitivity of the detector is given by

$$S = \eta \lambda q / hc$$

then

$$n_e = S\Phi / q \qquad\qquad (6.4)$$

For a CCD, the total number of charge carriers collected in integration period t_s is therefore given by

$$n_q = S\Phi t_s/q \qquad (6.5)$$

See Horowitz, P. and Hill, W. *The Art of Electronics* (Cambridge University Press, 1980) pages 286–91 or Jones, B.K. *Electronics for Experimentation and Research* (Prentice-Hall, 1986), Chapter 4.

where Φ is the radiant flux falling on the detector. As is well known, the shot noise generated in a detector is related to the square root of the detected signal and is given, therefore by $(S\Phi t_s/q)^{1/2}$. To this must be added any noise due to dark current under a single r.m.s. noise component, n_d. Thus the total noise component per *pixel* is $(n_d^2 + n_q^2)^{1/2}$. The signal-to-noise ratio is thus given by,

$$S/N = n_q t_s/(n_d^2 + n_q^2)^{1/2} \qquad (6.6)$$

When adding random, incoherent noise components, it is the square of the noise component which is added.

where t_s is the sampling time per pixel. We can see that since the signal increases with pixel area and the noise increases with the square root of the area, the signal-to-noise ratio increases with picture area.

Exercise 6.3 Calculate the signal-to-noise ratio of a CCD array of 294×604 pixels, each of area $10 \ \mu m \times 15.6 \ \mu m$, when illuminated by light of irradiance $10 \ W \ m^{-2}$. Assume the sensitivity of the CCD to be $50 \ mA \ W^{-1}$ and that dark current and detector noise are negligible.

The Laser-Scanning Camera

See Rosing, B. British Patent No. 27570 (1908).

The idea of a camera based on a *flying-spot* of light being scanned across a subject can be dated at least as far back as the issue of a patent in the early part of this century. Baird's original TV system also had elements in common with the scanning camera. Some recent variations on the scanning-light principle include supermarket checkouts, laser printers and scanning microscopes to name but a few.

Applications of scanning laser systems are dealt with by Beiser, L. in *Laser Applications* Vol. 2, ed. M. Ross, (Academic Press, 1974).

In the *Laser-Scanning Camera*, a bright spot of laser light is scanned across the field of view in a precise and well-defined pattern, known as a *raster* scan. The reflected light is synchronously detected by a photodetector and the electrical signal used to directly modulate the brightness of the raster-scanned spot on a TV monitor. The advantages of the technique are that no additional lighting is required, there is high sensitivity, large depth-of-field and the capability to *zoom* in and out of the image. As we have seen in the previous section, conventional TV cameras operate on the principle that broad-band light reflected from a scene is imaged onto a photosensitive surface such as a vidicon tube or solid-state array. The photon-generated signal is raster scanned by an electron beam from which a synchronously scanned spot on a TV screen is modulated to produce a recognizable image.

For principles of television see Zworgkin, V.K. and Morton, G.A. *Television: the Electronics of Image Transmission in Colour and Monochrome* (Wiley, 1954).

The impetus for the development of the laser-scanning camera came from the need of the nuclear-power industry to regularly inspect the interior of reactor installations in order to maintain reliable, efficient and safe operation of the generating plant. Because of the hostile environment and difficult access to the internal areas of the reactor extensive use is made of remote TV systems, both as means of direct visual inspection and also to guide manipulator arms to the inspection site. Conventional TV systems, however, do not perform well in the extremely hot and highly radioactive conditions found inside reactors. Even the small CCD or diode-array cameras, so perfect in many other ways for such applications cannot survive the high radiation levels. Additionally, all such cameras suffer from poor sensitivity at low light levels and possess only moderate resolution.

Camera Design

The basic layout of the scanning camera is shown in Fig. 6.16. The use of a laser as light source provides a beam of high intensity which is easy to focus to small spots, even over large distances, and can be scanned over the scene of interest. In the version of the camera discussed here a 7.5 mW HeNe laser is used although systems based around a laser diode are being developed.

In the development of a prototype laser-scanning camera the design was based around a HeNe laser. Put yourself in the position of the design team and give reasons for your choice of laser? What factors should be taken into account if the gas laser is to be substituted by a laser diode?

Exercise 6.4

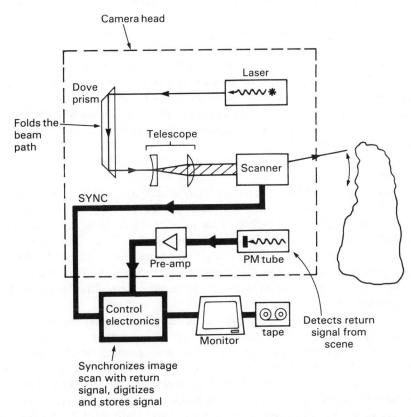

Fig. 6.16 Schematic diagram of the laser-scanning camera (reproduced by permission of CEGB).

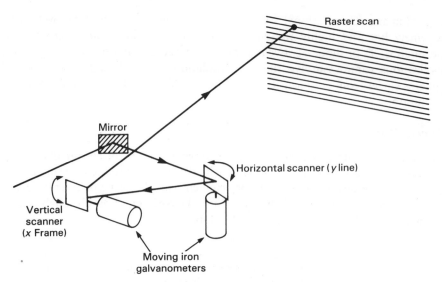

Fig. 6.17 Orthogonal scanning system.

This type of mounting is known as orthogonal mounting.

Alternative methods of scanning laser beams are dealt with in some detail by Beiser, L. Laser Scanning Systems in *Laser Applications* Vol. 2, ed. M. Ross (Academic Press, 1974).

See Chapter 4 for a discussion of photomultipliers.

The beam scan is often accomplished by bouncing it between two mirrors mounted on oscillating galvanometers with their axes of rotation mutually perpendicular to each other (Fig. 6.17). The beam is scanned across the field of view in a raster pattern by incremental movement of one or other of the two mirrors. The maximum angle of scan in this system is 40° with a scan rate of 1 kHz. Although *acousto-optic* and *electro-optic* deflectors are available which permit scanning rates up to several megaHertz, the angle of scan is only a few degrees, thereby considerably reducing the field of view. Such deflectors are also much more expensive than galvanometer scanned systems. Light reflected from the scene is detected by an 11-stage PM tube with an S-20 photocathode and the output fed to the processing electronics.

The resolution of the final image is dependent on both the spot size to which the scanned beam is focused and the image contrast. Furthermore, the smaller the spot size at the target the greater is the irradiance delivered to it and, consequently, the greater will be the intensity of the returned signal. To meet this requirement for a small spot size the initial divergence of the laser beam has to be reduced. This is accomplished by expanding the beam diameter by some factor over its nominal diameter of 0.8 mm and refocusing through the scanner to the object. In this case expanding the beam by a factor of three reduces the divergence from 0.6 mrad to 0.2 mrad. Focusing the beam is carried out by moving the positive lens in the telescope combination.

Developing any instrument for a specific industrial application demands that the environmental conditions relating to its use be taken into account during its design. For instance, in the applications for which the scanning camera is intended, its size and shape are of crucial importance as are its resistance to heat, radiation and stray light. To reduce the camera dimensions sufficiently for it to be easily manipulated in the cramped conditions of a reactor, beam-folding optics are used in the form of a *dove prism* which bends the beam through 180°. Additionally, because this camera is designed for in-reactor use the PM tube is encased in a 6 mm thick water-

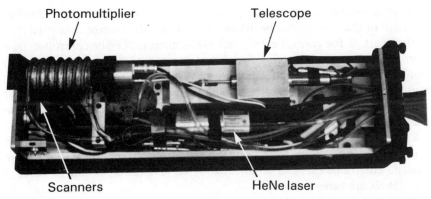

Fig. 6.18   Laser-scanning camera head (reproduced by permission of CEGB).

cooled copper housing both to protect it from the intense heat and to act as an absorber of beta and gamma rays which induce noise in the tube. Although the camera is intended to operate in a light-free environment, on occasion it will be operated in the presence of auxiliary lighting. To shield the camera from such ambient light, interference filters with a narrow band-pass of about 10 to 50 nm centred around the laser wavelength, are placed in front of the lens.

A photograph of the camera head is shown in Fig. 6.18.

Control Electronics

The control electronics are shown in block form in Fig. 6.19. Before passing to the control, the electrical signal from the photomultiplier is first amplified to around 1 volt amplitude and filtered to remove any components above 100 kHz. In this way the signal has sufficient power to be driven along a 50 m length of coaxial cable and still provide a large enough signal for the rest of the electronics. By keeping the

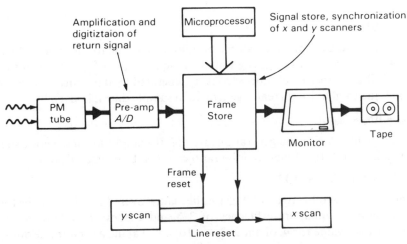

Fig. 6.19   Control electronics for scanning camera (reproduced by permission of CEGB).

bulk of the electronics remote from the camera head, size is once again minimized and control of the camera can be maintained outside the reactor area itself.

At the heart of the control electronics is the microprocessor-controlled *frame grabber*. From here the triggering signals which generate the frame and line scans go out to the y and x scanners, respectively. Each frame consists of 512×512 measurement points. The line scan is initiated by a clock pulse to the x scanner. The number of clock pulses is monitored by a digital counter and used in conjunction with a D-to-A converter to generate a voltage ramp to the scanner. After a count of 512, a line-reset pulse restarts the counter for the next line. Similarly, the frame scan is initiated by a pulse to the y scanner and a corresponding voltage ramp generated. The frame rate can be adjusted by varying the clock rate at which the trigger signals are generated.

The returned signal from the camera is digitized and stored as a 8-bit, 512×512 video frame in the frame store. The contents of the frame store are converted to composite video format and fed to a TV display at normal TV rates.

Signal-to-Noise Ratio

A consideration of the signal-to-noise ratio of the camera will enable us to estimate the range of operation of the instrument. The rate of photon arrival at the target, for a laser flux Φ is given by

$$n_{\text{inc}} = \Phi\lambda/hc$$

Accordingly if the target is a diffuse reflector of reflectivity R, the light reflected back to a photocathode at a distance z from the target is,

$$n_{\text{ref}} = R\Phi\lambda/hc(2\pi z^2)$$

The term $2\pi z^2$ indicates that the reflected light will be re-radiated into a hemisphere of radius z. Hence, the number of photoelectrons generated by a photocathode of quantum efficiency η and area A is

$$n_{\text{e}} = \eta R\Phi A\lambda/hc(2\pi z^2) \tag{6.7}$$

The signal current generated in each picture element (*pixel*) in sampling time t_s is therefore,

$$i_{\text{e}} = n_{\text{e}} t_s$$

Assuming that the dominant source of noise is that due to shot noise, then the noise generated by a current i_e is given as the square root of i_e. Adding a typical figure for the number of photoelectrons n_d generated by dark current of around 1000 s^{-1}, the total noise current i_n generated in sample time t_s is given by

Noise signals separately appear as square-root functions. See Horowitz, P. and Hill, H. *The Art of Electronics* (Cambridge University Press, 1980) pages 286–91 or Jones, B.K. *Electronics for Experimentation and Research* (Prentice-Hall, 1986), Chapter 4.

$$i_{\text{n}} = [(n_{\text{e}} + n_{\text{d}})t_s]^{1/2}$$

At such high photo-electron generation rates it is the amplitude of the noise signals which add. Hence, the signal-to-noise ratio can then be estimated from

$$S/N = n_{\text{e}} t_s^{1/2}/(n_{\text{e}} + n_{\text{d}})^{1/2} \tag{6.8}$$

Assuming a typical reflectivity of 0.05 for the oxide-covered steel likely to be found inside reactors, a quantum efficiency of 3% and a photocathode diameter of 23 mm, yields a value for n_e of about $2.4 \times 10^9 z^{-2}$. Taking a total frame duration of, for example 8 s, corresponding to an exposure per pixel of

$$8 \text{ s}/(512 \times 512)$$

and a target distance of 5 m, the signal-to-noise ratio can be estimated to be about 50:1. This compares favourably with what might be expected from conventional CCTV cameras in similar conditions.

The laser-scanning camera is now in routine use in many of the UK's nuclear reactors.

High-Resolution Visual Inspection Using Holography

Holography is the last, but not least, of the applications of optoelectronics which we will discuss. It is the application which has the most dramatic visual appeal and has caught the imagination of the general public more than any other. No one who has seen a hologram recreated by laser light can fail to be amazed by the realism of an image which floats in space in front of the observer. Like most of the applications we have already looked at, holography originated well before the invention of the laser, but again had to await the arrival of this device before its engineering application could come to fruition. It has now found widespread industrial acceptance in such diverse areas as vibration analysis, image processing, particle sizing and visual inspection, whilst holographically-produced optical elements are essential components in aircraft head-up displays, supermarket scanners and optical computers.

The pioneering papers in holography are those by the founding father, Gabor, D. *Nature* Vol. 161, pages 777–8, 1948 and the men who turned it into a practical reality, Leith, E. and Upatnieks, J. *Journal of the Optical Society of America*, Vol. 53, pages 1123–30, 1962.

The uniqueness of holography comes from its ability to recreate, from a *two-dimensional* recording medium, a *three-dimensional* image which is to all intents and purposes *optically indistinguishable* from the original scene. Holography accomplishes this by recording, not an image of the object itself, but, the wavefront emanating from it. The resulting hologram bears no physical resemblance to the original object yet contains, in the form of an interference field, all the optical information about it. Subsequent reconstruction of a recognizable image from the hologram by illuminating it with laser light reproduces, remote from the original scene, a life-size, three-dimensional image of high resolution and low in optical aberrations.

Because of these characteristics, holography lends itself to a potentially powerful method of visual inspection, now known as *hologrammetry*, in which high-resolution measurement can be performed on holographic images reconstructed remote from the original scene. This technique is particularly useful in those areas where the inspection has to be carried out in a hazardous and inaccessible environment. Some notable examples are in-reactor inspection of fuel elements for the nuclear-power generating industry and underwater inspection of offshore oil and gas installations.

We will make use of holographic visual inspection as a vehicle with which to understand the basis of holography and appreciate its unique character. We will also discuss the requirements of a holographic camera for on-site visual inspection.

Holographic Recording

To understand the unique properties of holography and to see how the hologram can be used as the basis of visual inspection it helps if we first compare it with the photographic process. A scene illuminated with broad-band light from an external light source, such as the sun or flashtube, is shown in Fig. 6.20. Light reflected from the object is collected by a lens and imaged onto a detector such as photo-

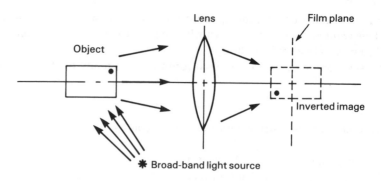

Fig. 6.20 Photographic process.

See Chapter 1, eqn (2.1).

graphic film, a TV tube or solid-state array. We have already seen that to mathematically describe the distribution of light in a given plane we need to specify both its *amplitude* and *phase*. Detectors such as those mentioned above, the so-called *square-law* detectors, are sensitive only to the square of the electric field distribution, and hence, record the distribution of light as a two-dimensional variation of light intensity across the image plane. Phase information is lost. Since it is this phase relation between the different rays of light as they meet at the film which tells us the relative time history of each ray, the loss of this component compromises our ability to record the three-dimensionality and parallax of a scene.

Stereo-photography retains the three-dimensionality of a scene within the limited depth of field of the lens system, but still loses the parallax information.

    In contrast, Fig. 6.21 shows the basic requirements of holographic recording. Some differences are immediately apparent between this procedure and that relating to photography. In general, no lenses are used in the image formation process: as we have already said, it is the *wavefront* of the light reflected from the object which is incident on the recording medium not a focused image of it. The lenses in holography act as beam expanders which are used to control the respective illumination of object and film. Next, two beams of light, rather than one, are used in the recording process. The intensity of the illuminating light source is split into two, each part of which travels a different path. One part, the *reference wave*, is expanded by a lens and illuminates the holographic recording medium directly. The second part, the *object wave*, is also expanded and directed onto the object itself. The wavefront reflected from the object travels towards the recording

This is not the case in all forms of holography, but is true for the particular case we are interested in.

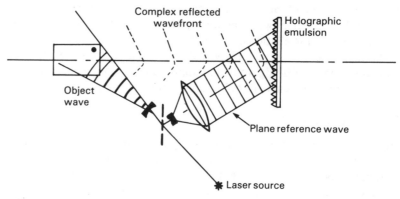

Fig. 6.21 Hologram recording process.

medium where it combines with the reference beam. The way in which the two beams combine is governed by the third significant difference between holography and photography, namely, that a laser is used as the illuminating source rather than broad-band light. Lasers, we have already seen, are sources of coherent light. Since the object and reference beams are derived from the same laser, the two beams will be mutually coherent, and can therefore *interfere* to form a two-dimensional pattern of light intensity in space. It is the formation and recording of this interference pattern which holds the key to the holographic process.

A word, briefly, about the recording medium. In general, the holographic interference field is captured on photographic film. Holographic film differs from the film used in ordinary photography only in that the grains of silver halide are the order of a few nanometres across as compared to micrometers. Such film is very insensitive to light but has the capacity to record the fine detail inherent in an interference field. The exposed film is chemically processed in a similar, but somewhat more elaborate, way to ordinary film to render the holographic interference permanent. There is a fine balance to be aimed for between image brightness and resolution. This stage is crucial if high-resolution measurements are to be made from the hologram.

Image Formation

A look at the process of optical interference in more detail will show how recording such a pattern helps us reconstruct an image of an object. Consider the simple situation of interference between two plane waves of monochromatic, coherent light, $\mathscr{E}_0$ and $\mathscr{E}_r$, meeting in space at an angle β (Fig. 6.22). Because the two beams are mutually coherent, the peaks and troughs will by superposition produce the characteristic two-beam interference pattern of alternating dark and light parallel lines. The pitch of the interference pattern depends on the wavelength of light and the angle between the two beams and is the order of half a wavelength.

If we record the interference field on fine-grain photographic film, such an interference pattern will act as a sinusoidal diffraction grating. Illuminating the grating with one of the two beams used in its construction, say $\mathscr{E}_r$, will generate *three* beams

Coherence was discussed in Chapters 2 and 3.

Interference was briefly discussed in Chapter 1.

Other media such as thermoplastic film, photochromic materials, non-linear optical crystals and dichromated gelatin have very specialized uses and will not be discussed here.

Typical sensitivity is around a few milliJoules per square metre.

For further details on holographic materials and their processing, see Hariharan, P. *Holography* (Oxford University Press, 1984) or Collier, R.J., Burckhardt, O.B. and Lin, L.H., *Optical Holography* (Academic Press, 1971).

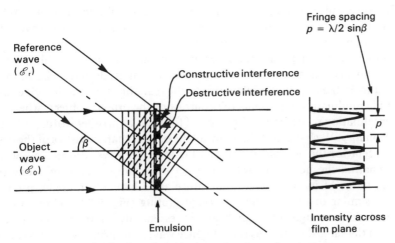

Fig. 6.22   Interference between two plane waves.

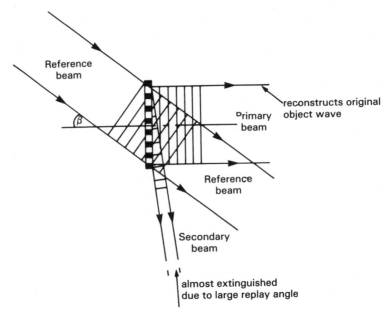

Fig. 6.23   Reconstruction from plane grating.

See Hecht, E. and Zajac, A. *Optics* (Addison-Wesley) for a mathematical justification.

There is an obvious analogy here between amplitude and frequency modulation of communications theory, except here we have amplitude and phase modulation combined. See Appendix C.

on the opposite side of the grating (Fig. 6.23). One beam, the *straight-through* beam, passes through the film in a direct line with the illuminating wavefront, $\mathcal{E}_r$. A second beam propagates at an angle $+\beta$ to $\mathcal{E}_r$ in line with the *apparent* direction of $\mathcal{E}_0$ and represents a replica of this wave. This is known as the *primary wave*. The third beam, known as the *secondary wave*, propagates at an angle $-\beta$ to the straight-through beam and contains elements of both $\mathcal{E}_0$ and $\mathcal{E}_r$. We have therefore produced an image of one wave by illuminating the two-dimensional grating with the other.

Now imagine the situation where one of the beams taking part in the interference process is not a parallel beam, but is the reflected light from some illuminated object, $\mathcal{E}_0$. Then that beam will possess a complex wavefront representing the amplitude and phase distribution of light in the object. The interference pattern now recorded will also be a complex field of varying amplitude and phase. What we have done is modulate the amplitude and phase of a fixed reference wave with the distribution of light received from the scene.

Again, by capturing this interference field on photographic film, we have recorded on a two-dimensional medium both the amplitude and phase of the light reflected from the object. To relate this to the real object we can imagine that the amplitude modulation represents the contrast of the original object, whereas, the phase modulation relates to the direction of the rays, or in other words, to its three-dimensionality. We have stored on film all the optical characteristics of the original scene, except for the notable exception of colour. We have recorded a hologram.

Illuminating the hologram with an exact duplicate of the original reference wave, in terms of its wavelength, curvature and illumination angle, produces a response similar to that of the sinusoidal grating (Fig. 6.24). Three waves are generated. The straight-through beam of essentially uniform irradiance propagates in the same direction as the reference wave. The primary wave is generated at an angle $+\beta$ to the reference and replicates the original object wave. If the eye

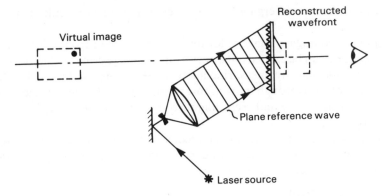

Fig. 6.24   Reconstruction of virtual image.

intercepts this wave it cannot distinguish between it and that from the original, even though there is an obvious time lapse in history between the two waves. The eye sees a life-size image of the original scene which appears to be located behind the photographic plate. The eye has formed a *virtual image* of the original. The secondary wave proceeds at an angle $-\beta$ to the reference wave and contains components of both object and reference waves. In practice, because β is usually in excess of 30° or so, the secondary wave is extinguished by internal reflection in the hologram plate.

For purposes of visual inspection, however, creation of the virtual image is not the most suitable form of holographic reconstruction. It so happens that if we turn the plate around and illuminate it from *behind* with a wave which is the exact conjugate of the reference beam, a conjugate image will be located in *real space* in front of the plate (Fig. 6.25). The image so created is optically identical to the original save that it appears to be reversed left to right and back to front. It is this real, *psuedoscopic* image which forms the basis of a method of visual inspection.

Principles of Visual Inspection

The utilization of holography in visual inspection, relies on the creation and optimization of the real image of a scene or object. The real, or more correctly, the *conjugate image*, is formed and reconstructed as shown in Fig. 6.25. A parallel reference beam is often used in recording the hologram, since reconstruction then is a simple case of turning the film around. If a diverging reference beam was used

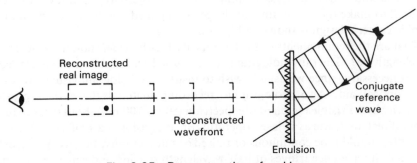

Fig. 6.25   Reconstruction of real image.

137

in recording, then a converging beam of identical curvature would be needed in reconstruction.

Because the conjugate image is located in real space in front of the observer, visual inspection can be carried out directly on this image using all the conventional tools of the trade, namely, measuring microscopes, photography and TV. Optical sections can be taken through the resulting reconstruction by merely placing a piece of film across the image and recording directly, *without the need for any lenses* (Fig. 6.25).

A further advantage of real-image reconstruction is that image resolution is limited only by the quality of the reproduced hologram. The limiting resolution r in line-pairs per millimetre is given by

$$r = 1/(1.22z/D)10^3$$

where z is the object-to-film distance, λ is the wavelength of the laser light and D is the hologram diameter. This equation is the standard relation of optics which defines the resolving power of a lens. A lens and a hologram of equivalent diameters will produce the same theoretical resolution. Because of its reduced susceptibility to optical aberrations, the hologram will produce the more highly resolved image. Conversely, the resolution of a virtual-image hologram is limited by the effective aperture of the viewing system.

This concept is sometimes hard to accept without seeing it. An image is actually formed in space in front of the observer on which you can perform all optical tests as if it were the original subject.

The speckle nature of the laser-reconstructed image, however, reduces the ideal resolution by a factor of two to three.

Exercise 6.5

Estimate the ideal resolution of a scene recorded with ruby-laser light on a standard 12.5 cm × 10 cm holographic plate with an object-to-film distance of 0.5 m.

[378 lp mm⁻¹]

It should be realized that the resolution obtained in this way assumes that we reconstruct with an exact conjugate of the reference beam, that the reconstruction wavelength matches that of the recording beam and that the emulsion is infinitesimally thin. Fig. 6.26 shows a photograph taken from a real-image reconstruction of a hologram. Parts of the object which are seen to be out of focus in this figure can be brought into focus simply by moving the imaging system, as if we were viewing an actual object.

Holographic Camera

The requirements for a holographic camera which could be used in the field to record holograms for later real-image reconstruction are fairly demanding. The camera should consist of a laser optimized for holography, optical components required to make the hologram and the recording medium itself. A typical holographic camera design is shown in Fig. 6.27.

What are the requirements of the laser itself? Firstly, what should the operating wavelength be? Generally holograms are recorded and replayed in visible light. This is necessary, since we usually wish to inspect the hologram visually or using some optical aid. Also, holographic recording materials have their greatest sensitivity in the visible region of the spectrum (500 to 700 nm). Both these requirements direct us towards a laser operating in the visible part of the spectrum. Secondly, should it be continuous or pulsed? In holography, because we record a pattern of interference fringes of half-wavelength spacing, movement in *any* part

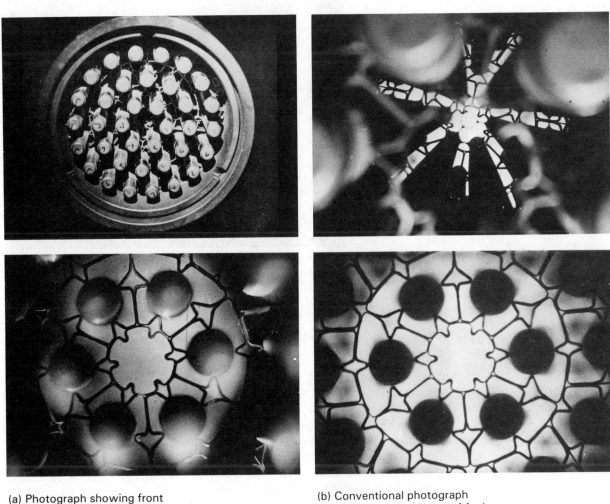

(a) Photograph showing front grid of AGR fuel element

(b) Conventional photograph focused on central brace of fuel element (approx $\frac{1}{2}$ m from front) showing perspective distortion, limited depth of focus and obscuration of field of view

(c) Photograph taken from real image reconstruction of hologram of fuel element. Photograph shows front brace

(d) Photograph taken from real image reconstruction of hologram of fuel element. Photograph shows middle brace (approx $\frac{1}{2}$ m shown). Note absence of distortion, depth of field limitations and obscuration

Fig. 6.26 Comparison between photographic visual inspection and holographic visual inspection.

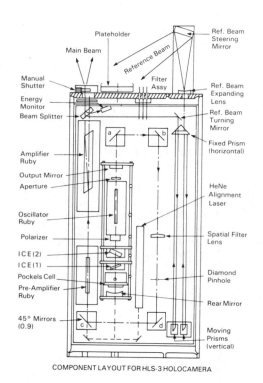

COMPONENT LAYOUT FOR HLS-3 HOLOCAMERA

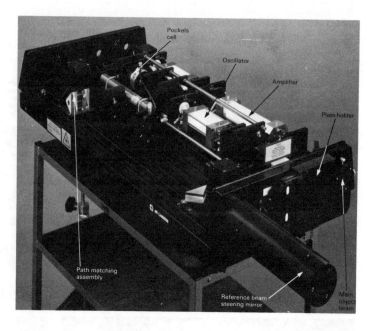

Fig. 6.27   Holographic camera.

of the holographic set-up of a fraction of a wavelength will wipe out the interference pattern and neighbouring fringes will merge into one another. Holography using a continuous-wave laser dictates the use of massive, vibration-isolated optical tables. The way around this, in the field, is to use a laser with a pulse duration of 50 ns or less. Over such a pulse duration, any object movement is effectively frozen. Thirdly, what output is needed? Holographic materials are basically insensitive to light, typically requiring a few millijoules per square metre at the film; this is because of their inherently fine grain. To counter this the laser should have a high output energy. Furthermore, high energy is needed to illuminate large scenes of low contrast which are likely to be encountered in industry.

Exercise 6.6   From what you have already learned about lasers and your understanding of holography, choose a laser to produce high-resolution holograms of large industrial objects. Give some reasons for your choice.

If you have thought about the above exercise, you should have arrived at the pulsed ruby laser as being the best compromise for the specific task in hand. A typical specification for such a laser would be that it deliver 1 J over a 50 ns, Q-switched pulse.

These are not the only requirements necessary for high-quality holograms of large objects. To ensure that the entire depth of scene is recorded, no part of the object should fall beyond the coherence length of the laser. If it did, light reflected from these parts of the object would not interfere with the reference beam, and consequently, would not be recorded on the hologram. Longest coherence lengths are obtained with the laser operating in a single longitudinal mode. Since longitudinal-mode control is dependent on cavity length, a temperature-stabilized etalon is normally incorporated to restrict the number of oscillating modes which are under the gain curve. Fine control is complex. Suffice to say here that the laser is designed with the required mode-controlling optics incorporated. Correctly designed the laser will produce coherence lengths in excess of one metre. Because maximum visibility of interference fringes will only be obtained when the path lengths are exactly matched, it is important that the camera incorporates path-length matching. This is accomplished using the path system shown in Fig. 6.27.

See Chapters 2 and 3.

Finally, to ensure even illumination of the scene we want the laser to operate in a single transverse mode. Transverse-mode control is a function of mirror separation, curvature and cavity width. Incorporation of an adjustable aperture in the cavity helps to ensure that the laser operates in the TEM_{00} mode.

Summary

In the last few pages we have outlined some of the systems which typify the application of optoelectronics in modern engineering. The systems chosen were drawn from actual industrial case studies to highlight the diversity of application and give an indication of the design criteria applied in the selection of the most appropriate components.

Problems

6.1 What factors have to be taken into account in the design of a repeaterless digital communications channel based on fibre-optics? You should consider, in turn, the selection of fibre, souce and detector.

6.2 What parameters should be taken into account in the design of an atmospheric line-of-sight communications link between two adjacent buildings?

6.3 If the refractive index of the core of a step-index fibre is 1.53, determine the refractive index of the cladding if light is to be transmitted along a 5 km length of the fibre at a data transmission rate of 8 Mbit s^{-1}.

6.4 It is intended to use an LED and a silicon photodetector as the basis of a smoke detector in a long, narrow corridor. Placing the LED at one end of the corridor and the detector at the other, smoke will be detected when the light falling on the detector falls below a pre-set value. Using the data in Tables 4.1–4.4, discuss the most suitable choice of photodetector. Make reasonable assumptions about the environment or the system which you find necessary.

6.5 A pulsed ruby laser is to be used as the basis of a lunar rangefinder. If a 0.5 m × 0.5 m array of mirrors is placed on the surface of the moon, estimate the number of photons which could be collected by a 2.5 m diameter telescope back on earth. Assume that the laser has an output energy of 1 J, a pulse duration of 30 ns, an output beam diameter of 200 mm and a divergence (after recollimating optics) of 0.03 mrad. The approximate earth to moon distance is

400 000 km. Think carefully about any assumptions you may have to make.

6.6 Assuming that no pulse shaping optics or other means of improving the resolution are available, what is the maximum possible accuracy on the lunar range measurement of the previous example?

6.7 A simple method of measuring the turbidity of a gas or liquid involves containing the medium in a glass cell and irradiating with a beam of light. On the opposite side of the cell a photodetector monitors the transmitted light. Discuss the parameters involved in the practical implementation of the technique. What factors influence your choice of source and detector?

6.8 If the exhaust outlet of a jet engine is at a temperature of 550 °C and has an effective diameter of 400 mm, estimate the maximum range at which a guided missile fitted with a heat-sensing detector will lock-on to the aircraft. The minimum radiant flux which the sensor will respond to is 1 μW and its sensing area is 100 mm diameter. (Hint: assume the spatial profile of radiation emitted from the exhaust to be Lambertian. Refer also to Appendix A on black-body radiation.)

Appendix A

Black-Body Radiation

When atoms are close together, as in a dense gas or solid, the energy structure associated with the individual atoms influence one another. The energy levels overlap and split in proportion to the number of atoms in the medium. Any emitted radiation will correspond to transitions from bands of energy levels and will consist of many closely spaced wavelengths giving rise to a continuous spectrum. This is the collective behaviour of many interacting atoms of different species, rather than the characteristic behaviour of individual atoms of a particular element.

The radiation profile emitted from such a body can be described in terms of an ideal thermal radiator: *the black body*. An ideal radiator is one which emits as much energy as it absorbs. The overall spectral distribution is given by the *Planck black-body* equation which relates the radiant energy emitted by an ideal radiator to its temperature, as follows,

$$M_e(\lambda) = \frac{2\pi hc^2}{\lambda^s} \left[\frac{1}{\exp(hc/\lambda kT - 1)} \right] \tag{A.1}$$

The above relation describes the radiant flux emitted per unit surface area per unit wavelength interval from the surface of a source into a 2π solid angle. The equation as it stands applies to unpolarized light: for linearly polarized light we should divide by 2.

A family of spectral radiant-emittance curves for a black body as a function of emitted wavelength and temperature is shown in Fig. A.1. Some important points should be noted about these curves:

(i) the total radiant emittance at a given temperature is equal to the area under the curve;
(ii) the emitted flux increases as the temperature of the body increases: as the body gets hotter it emits more energy;
(iii) the peak of the curve shifts to lower wavelengths, or higher frequencies, as the temperature increases.

This relation was first developed by the famous Danish physicist Max Planck in 1900 and was a milestone in the understanding of the nature of light.

As a piece of steel is heated up its colour changes from dull red through orange to *white-hot*.

The total flux emitted from the source is obtained by integrating $M_e(\lambda)$ over all wavelengths up to λ_{max}, and is given by

$$M_e = \sigma e T^4 \tag{A.2}$$

where σ is the Stefan constant ($\sigma = 56.7$ nW m^{-2} K^{-4}) and e is a factor, called the emissivity of the surface, which varies between zero for no radiant emission and unity for a black body.

Estimate the radiant emittance of a black body at a temperature of 300 K.   *[460 W m^{-2}].*   **Exercise A.1**

143

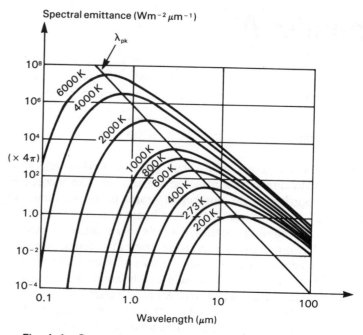

Fig. A.1 Spectral emittance isotherms for a black body.

The maximum radiant power which can be emitted from any source at a given temperature is described by eqn (A.2); the radiant emittance from a line source at the same temperature can never be greater than this value.

Exercise A.2   By making sensible assumptions about the surface area of a human body and its radiant temperature, calculate the total flux emitted.

[≈ 900 W]

For a more exhaustive treatment of black-body radiation the reader should consult a text such as Born, M. and Wolf, E. *Principles of Optics* (Pergamon, 1980).

Finally, for each constant temperature curve, known as an *isotherm*, the peak wavelength is obtained by differentiating eqn (A.1) with respect to λ and equating to zero. Hence,

$$\lambda_{pk} T = hc/4.97k \tag{A.3}$$

Exercise A.3   If you wanted to take thermal pictures of the human body in the previous example, at what wavelength should the thermal camera have its peak sensitivity?

[10 μm]

In developing the black-body relation, Planck had to assume that the energy distribution within the source was not a continuous function of frequency as assumed up until then, but existed in discrete quanta of energy known as photons. On this basis, the energy was distributed amongst permitted modes of oscillation within the source volume. The existence of these modes is analagous to the permitted modes of vibration of a stretched string or the modes of oscillation within an optical cavity as discussed in Chapter 2.

Appendix B

Principles of Laser Action

We have discussed, in the main text, the principal concepts of laser action from a primarily phenomenological standpoint. Now we can present some of the physics behind many of the arguments already voiced.

The Einstein Relation

The significance of stimulated emission as far as laser action is concerned can be seen by considering the following situation. An atomic system consisting of an upper and lower energy state of respective energies W_u and W_l and initially populated by N_u and N_l atoms per unit volume, respectively, is irradiated by light of frequency f_{ul} and spectral energy density $dW(f)/dV$ (Fig. B.1). Spectral energy density is the energy emitted, from an ideal source, per unit volume per unit frequency interval, and is defined by

$$\frac{dW(f)}{dV} = \frac{8\pi f^2}{c^3} \frac{hf}{\exp(hf/kT) - 1} \qquad (B.1)$$

This is the *Planck black-body* relation described in Appendix A, but restated here in a slightly different form. The SI units of $dW(f)/dV$ are J Hz^{-1} m^{-3}.

The relative population of each energy level depends on the interplay between three processes, namely:

Spontaneous and stimulated emission and absorption are discussed in Chapter 2.

(i) excitation of atoms from level l to level u by absorption of photons from the irradiating beam;
(ii) de-excitation of atoms from level u to level l by spontaneous emission of photons;
(iii) de-excitation of atoms from level u to level l by stimulated emission of photons.

The radiant flux involved in these processes depend upon the initial populations

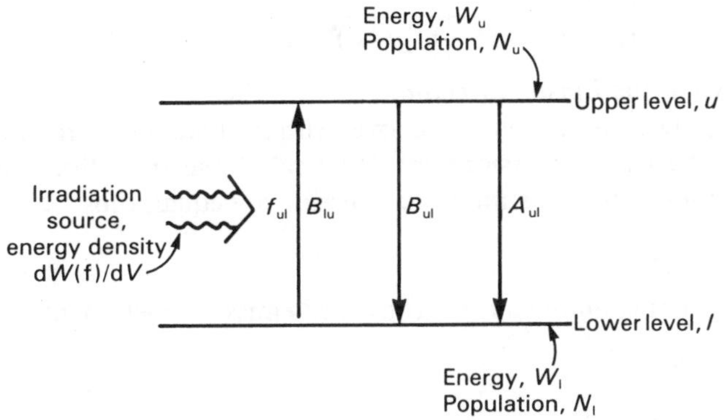

Fig. B.1 Irradiation of two-level atomic system.

of the levels, the energy density of the irradiating beam and the likelihood of a particular transition taking place. The likelihood of a transition occurring is described in terms of the *Einstein A and B coefficients*, A_{ul}, B_{ul} and B_{lu}, which refer respectively to spontaneous emission between u and l, stimulated emission between the same levels and absorption between l and u. The spontaneous-emission coefficient is itself related to the spontaneous lifetime t_{spont} of the transition by

$$A = 1/t_{spont} \tag{B.2}$$

The flux which is absorbed by the medium per unit volume in raising atoms from level l to level u is therefore

$$d\Phi_{abs}/dV = [B_{lu}N_l dW(f)/dV]hf_{lu} \tag{B.3}$$

The SI units of B are m³ J⁻¹ s⁻².

where f_{lu} is the frequency of irradiation.

The radiant flux emitted by spontaneous transitions from level u to level l, is given by,

$$d\Phi_{spont}/dV = A_{ul}N_u hf_{ul} \tag{B.4}$$

The SI units of A are s⁻¹.

where f_{ul} is the frequency of the emitted photons.

Finally, the flux emitted by stimulated transitions from level u to level l, is given by

$$\frac{d\Phi_{stim}}{dV} = B_{ul}N_u \frac{dW(f)}{dV} hf_{ul} \tag{B.5}$$

For a system in thermal equilibrium with its surroundings, the flux absorbed must equal the total flux emitted and hence,

$$\frac{d\Phi_{abs}}{dV} = \frac{d\Phi_{spont}}{dV} + \frac{d\Phi_{stim}}{dV} \tag{B.6}$$

or

$$B_{lu}N_l \frac{dW(f)}{dV} = A_{ul}N_u + B_{ul}N_u \frac{dW(f)}{dV} \tag{B.7}$$

Try these steps for yourself.

After some mathematical juggling and substituting the Boltzmann relation (2.7) into eqn (B.7), we arrive at the following expression for the spectral energy density of the illuminating source,

$$\frac{dW(f)}{dV} = \frac{A_{ul}/B_{ul}}{B_{lu}/B_{ul}} \frac{1}{\exp(-hf_{ul}/kT) - 1} \tag{B.8}$$

This is known as the *Einstein Relation*.

Comparing the Einstein relation above, with the black-body relation eqn (B.1), we see that the two expressions are identical if the transition coefficients of stimulated absorption and stimulated emission are equal, thus

The subscripts can now be dropped for convenience.

$$B_{lu} = B_{ul} = B \tag{B.9}$$

and if the ratio of spontaneous to stimulated emission coefficients is given by,

$$A/B = 8\pi hf^3/c^3 \tag{B.10}$$

eqn (B.8) now becomes

$$\frac{dW(f)}{dV} = \frac{A/B}{\exp(-hf_{ul}/kT) - 1} \qquad \text{(B.11)}$$

Spontaneous and Stimulated Emission

From eqns (B.4) and (B.5), we can show the relative importance of spontaneous and stimulated emission in an atomic system at equilibrium. The ratio of flux emitted by stimulated emission to that emitted by spontaneous emission is

$$\Phi_{stim}/\Phi_{spont} = B[dW(f)/dV]/A \qquad \text{(B.12)}$$
$$= 1/[\exp(hf/kT) - 1] \qquad \text{(B.13)}$$

For the emitted flux to be equal, that is for stimulated and spontaneous emission to be equally likely, eqn (B.13) should equal unity; hence, rearranging gives

$$f = (kT/h)\ln 2$$

or in terms of wavelength,

$$\lambda = \frac{ch}{kT\ln 2} \qquad \text{(B.14)}$$

Example B.1

Show that at room temperatures and at visible wavelengths, stimulated emission is not a dominant process in a radiant atomic system.

Solution: Substituting various temperatures into eqn (B.14) shows that at 300 K, the wavelength where stimulated and spontaneous flux are equal is about 69 μm. We need to be upwards of 30 000 K before stimulated emission begins to dominate in the visible region of the spectrum. Hence, at room temperatures, stimulated emission is not a dominant process. Stimulated emission is believed to occur naturally towards the core regions of the sun.

69 μm corresponds to microwave wavelengths.

Number of Photons in a Mode

From eqn (B.10), we can estimate how many photons there are in any given mode of oscillation. A study of black-body radiation tells us that when light is emitted from an ideal source only certain frequencies of oscillation are permitted, known as *modes*. From the black-body relation, the actual number of modes per unit volume in a given frequency interval is $8\pi f^2/c^3$, and hence, the radiation density can be written as,

$$dW(f)/dV = \tilde{N}(8\pi f^2/c^3)hf \qquad \text{(B.15)}$$

where $\tilde{N}$ is the number of photons in a single mode.

Comparing eqn (B.15) with the black-body relation shows that the number of photons in a single mode is given by,

$$\tilde{N} = 1/[\exp(hf/kT) - 1]$$

In other words, the number of photons in a mode equals the ratio of flux emitted by stimulated emission to that emitted by spontaneous emission. The important point to be noted here is that it is stimulated emission rather than spontaneous emission which increases the number of photons in a given mode. Hence, as we have already seen, if our aim is to amplify light we must promote stimulated emission at the expense of spontaneous emission.

Gain of a Laser Medium

We showed in Chapter 2 how the gain of a laser medium is given in terms of the small-signal gain coefficient as

$$g = g_i + (1/2L)\ln(1/R)$$

We can now show how g is related to the population inversion in the medium. Referring once again to Fig. B.1, energy is extracted from the irradiating beam during the process of absorption, as it traverses the medium, and is added to it by stimulated emission of photons. Since we know that it is stimulated emission which adds photons in phase with the incoming beam (spontaneous photons are emitted randomly in space and time), we can neglect spontaneous emission when discussing the conditions required to amplify light.

We can now write the change in flux per unit volume of the incident beam as it passes through the medium as the difference between the absorbed flux and emitted flux. Hence, from eqn (B.7) and neglecting spontaneous emission,

$$\frac{d\Phi}{dV} = hfBN_{inv}\frac{dW}{dV} \tag{B.16}$$

where $f = f_{lu} = f_{ul}$, $B = B_{lu} = B_{ul}$ and $N_{inv} = N_u - N_l$.

Returning now to some electromagnetic theory we may remind ourselves that the flux of an electromagnetic wave travelling through a medium of refractive index n is related to its total energy density dW/dV and its velocity of propagation in free space c by

$$\Phi = \frac{ca}{n}\frac{dW}{dV} \tag{B.17}$$

See, for example, Compton, A.J. *Basic Electromagnetism* (Van Nostrand Reinhold,   ).

where a is the cross-sectional area of the beam perpendicular to its direction of travel. Thus, the change in flux can be written as

$$\frac{d\Phi}{dV} = \frac{\Phi N_{inv}Bhfn}{ca} \tag{B.18}$$

or

$$\frac{1}{\Phi}\frac{d\Phi}{dL} = \frac{N_{inv}Bhfn}{c} \tag{B.19}$$

The term $d\Phi/\Phi dL$ we may recognize as the small-signal gain coefficient g. Hence, we obtain the threshold condition for laser action as

$$N_{th}Bhfn/c = g \tag{B.20}$$

or, by rearranging,

$$N_{th} = (c/Bhfn)g \tag{B.21}$$

where N_{th} is the population inversion at threshold. From eqns (B.2) and (B.10) we can replace the stimulated emission coefficient B by the spontaneous lifetime of the transition. Hence,

$$B = c^3/8\pi hf^3 n^3 t_{spont} \tag{B.22}$$

Additionally we have to multiply the right-hand side of eqn (B.21) by the spectral linewidth Δf of the transition to account for the finite width of the line. Including these factors in eqn (B.21), we arrive at the following expression for the population inversion required to reach threshold in a given atomic system,

See Chapter 2 for a discussion of linewidth.

$$N_{inv} = [8\pi f^2 t_{spont} n^2 \Delta f/c]g \tag{B.23}$$

Estimate the population inversion required to produce laser action in a crystal of ruby with the following parameters.

Worked Example B.1

Central lasing frequency: 432 THz.
Spectral linewidth: 330 GHz.
Spontaneous lifetime of laser transition: 3 ms.
Small-signal gain coefficient: 1 m^{-1} (including internal and mirror losses).
Refractive index: 1.5.

Solution: Substituting the above values into eqn (B.23) yields the required population inversion as

$$N_{inv} = 8\pi(432 \times 10^{12} \text{ Hz})^2 \times 3 \times 10^{-3} \text{ s} \times 1.5^2 \times 330 \times 10^9 \text{ Hz} \times$$
$$1 \text{ m}^{-1}/(300 \times 10^6 \text{ m/s})^2$$
$$= 116 \times 10^{21} \text{ m}^{-3}$$

Using the data given below, estimate the population inversion required to induce laser action in a Nd-YAG crystal. Assume that the total losses including internal losses and output coupling losses are 1 m^{-1}. Compare your answer with that calculated for ruby in Worked Example B.1.

Exercise B.1

[Ans: 3.1 $\times$ 10^{21} m^{-3}]

Central lasing frequency: 282 THz.
Central lasing wavelength: 1.0640 μm.
Linewidth: 0.8 nm (210 GHz).
Spontaneous lifetime of laser transition: 200 μs.
Refractive index: 1.82.

Appendix C

Theory of Holography

In Chapter 6 we outlined the concepts of holography on a purely pheno-menological basis. We can put those arguments on a more mathematical footing by considering the following analysis.

Consider a hologram formed by interference between a wavefront $\mathscr{E}_0$ reflected from an object and a reference wave $\mathscr{E}_r$ as shown in Fig. C.1. We assume that the two waves are linearly polarized in the same plane and that the interference pattern is recorded on flat, infinitesimally thin film. The object wavefront can be written in terms of the complex field,

$$\mathscr{E}_0 = a_0\exp(-i\phi) \tag{C.1}$$

where a_0 is the complex amplitude of the reflected wave and ϕ is the phase variation across the field. For simplicity, and to facilitate real-image reconstruction, we may represent the reference wave by a *plane wave* of uniform amplitude a_r meeting the object wave at an angle β to the holographic film. Hence,

$$\mathscr{E}_r = a_r\exp(ikx\sin\beta) \tag{C.2}$$

where k is the propagation constant, $2\pi/\lambda$. The spatial frequency of the reference wave is, therefore, represented by $\sin\beta/\lambda$.

The irradiance of the interference pattern recorded at the film is given by

$$\begin{aligned} E_h &= |\mathscr{E}_0 + \mathscr{E}_r|^2 \tag{C.3} \\ &= (\mathscr{E}_0 + \mathscr{E}_r)(\mathscr{E}_0^* + \mathscr{E}_r^*) \\ &= (\mathscr{E}_0\mathscr{E}_0^* + \mathscr{E}_r\mathscr{E}_r^* + \mathscr{E}_0\mathscr{E}_r^* + \mathscr{E}_r\mathscr{E}_0^*) \tag{C.4} \end{aligned}$$

The complex conjugates of the object and reference waves are given by

$$\mathscr{E}_0^* = a_0\exp(i\phi) \tag{C.5}$$

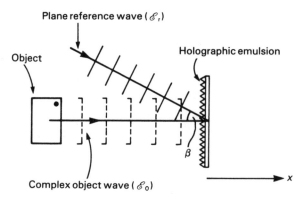

Fig. C.1   Recording of off-axis hologram.

150

and

$$\mathscr{E}_r^* = a_r \exp(-ikx \sin\beta) \tag{C.6}$$

Chemically processing the film will result in the recording of a hologram whose *amplitude transmittance* is linearly related to the irradiance at the film, by

$$t_a = bE_h \tag{C.7}$$

where b is a constant. Hence,

$$t_a = b(\mathscr{E}_0 \mathscr{E}_0^* + \mathscr{E}_r \mathscr{E}_r^* + \mathscr{E}_0 \mathscr{E}_r^* + \mathscr{E}_r \mathscr{E}_0^*) \tag{C.8}$$

Substituting for $\mathscr{E}_0$, $\mathscr{E}_0^*$, $\mathscr{E}_r$ and $\mathscr{E}_r^*$ yields the following expression:

$$
\begin{aligned}
t_a &= b\{[a_0\exp(-i\phi)a_0\exp(i\phi) \\
&\quad + a_r\exp(ikx\sin\beta)a_r\exp(-ikx\sin\beta)] \\
&\quad + [a_0\exp(-i\phi)a_r\exp(-ikx\sin\beta)] \\
&\quad + [a_r\exp(ikx\sin\beta)a_0\exp(i\phi)]\}
\end{aligned}
$$

$$
\begin{aligned}
&= b\{a_0^2 + a_r^2 + a_0\exp(-i\phi)a_r\exp(-ikx\sin\beta) \\
&\quad + [a_r\exp(ikx\sin\beta)a_0\exp(i\phi)]\}
\end{aligned} \tag{C.9}
$$

$$= b\{a_0^2 + a_r^2 + 2a_0 a_r \cos(kx\sin\beta - \phi) \tag{C.10}$$

The recorded interference pattern as shown in eqn (E.10) corresponds to a set of carrier fringes of spatial frequency $f_z = \sin\beta/\lambda$ due to the reference wave, modulated by the amplitude, a_0 and phase ϕ of the object wave. The depth of modulation of the fringes is given in terms of the *fringe visibility*, i.e.

$$V = 2a_0 a_r/[a_0^2 + a_r^2] \tag{C.11}$$

or, expressing a_0^2 and a_r^2 as the object and reference wave irradiances E_0 and E_r respectively,

$$V = 2(E_0 E_r)^{\frac{1}{2}}/(E_0 + E_r) \tag{C.12}$$

It is the fringe visibility which ultimately determines the brightness and contrast of the reconstructed image and so in hologram recording we want to optimize this factor as far as the capabilities of the film will allow.

Reconstruction of the image is accomplished by re-illuminating the recorded hologram by the original reference wave, $\mathscr{E}_r$. Hence the reconstructed wavefront is given by,

$$\mathscr{E}_c = t_a \mathscr{E}_r \tag{C.13}$$

Substituting for $\mathscr{E}_r$ from eqn (E.2) and for t_a from eqn (E.9), gives

$$
\begin{aligned}
\mathscr{E}_c &= b\{a_0^2 + a_r^2 + a_0\exp(-i\phi)a_r\exp(-ikx\sin\beta) \\
&\quad + a_r\exp(ikx\sin\beta)a_0\exp(i\phi)\}\, a_r\exp(ikx\sin\beta)
\end{aligned}
$$

$$
\begin{aligned}
&= b\{[a_0^2 + a_r^2]a_r\exp(ikx\sin\beta) + a_r^2 a_0\exp(-i\phi) \\
&\quad + a_r^2\exp(i2kx\sin\beta)a_0\exp(i\phi)\}
\end{aligned} \tag{C.14}
$$

The above equation contains three terms. The first, represents the reconstruction beam modified by a constant factor, and passes through the hologram in the same direction as the original reference wave (Fig. C.2). The second and third terms, respectively, represent the formation of primary and secondary images. The primary image contains the original object wave, $a_0\exp(-i\phi)$, multiplied by

151

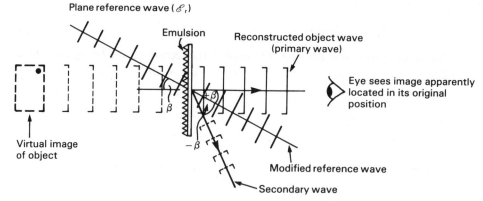

Fig. C.2 Reconstruction of virtual image.

the square of the reference-wave amplitude and travels in the same direction as the original object wave. The so-formed image is located behind the holographic plate and constitutes a virtual image of the scene. The secondary image contains elements of both the conjugate of the original object wave, $a_0\exp(i\phi)$, and the reference wave. This wave propagates at an angle defined by the exponential term, which corresponds to approximately $-\beta$ to the original reference wave. In practice, the finite thickness of the emulsion, about 10 μm, and reference beam angles in excess of 30° cause the secondary wave to be extinguished by internal reflections.

We can see, therefore, that by using an off-axis reference beam, the primary and secondary images are physically separated in space.

If, instead of illuminating with the original reference wave, we illuminate with its complex conjugate, the resulting wavefront becomes

$$\mathscr{E}_0^* = t_a \mathscr{E}_r^* \tag{C.15}$$
$$= b\{a_0^2 + a_r^2 + a_0\exp(-i\phi)a_r\exp(-ikx\sin\beta)$$
$$+ a_r\exp(ikx\sin\beta)a_0\exp(i\phi)\} \, a_r\exp(-ikx\sin\beta)$$

$$= b\{[a_0^2 + a_r^2]a_r\exp(-ikx\sin\beta)$$
$$+ a_r^2a_0\exp(-i\phi)\exp(-i2kx\sin\beta) + a_r^2a_0\exp(i\phi)\} \tag{C.16}$$

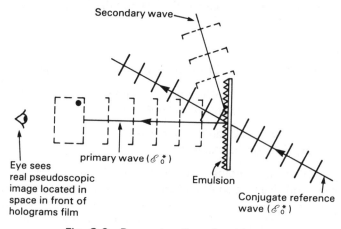

Fig. C.3 Reconstruction of real image.

152

Now we see that it is the conjugate wave which forms the primary image and proceeds in the direction of the original object wave. A *real image* of the object is formed on the opposite side of the hologram to the incident reference wave. This image, however, is formed back-to-front and reversed left to right. The image is known as a *pseudoscopic image* (Fig. C.3).

Appendix D

Physical Constants

Velocity of light (c)	299×10^6 m s^{-1}
Planck's constant (h)	663×10^{-36} J s
Fundamental electric charge (q)	160×10^{-21} C
Avogadro's number (N_A)	602×10^{24} kmol^{-1}
Electron mass (m_e)	911×10^{-33} kg
Proton mass (m_p)	1.67×10^{-27} kg
Electric constant of free space (ϵ_0)	8.85×10^{-12} F m^{-1}
Magnetic constant of free space (μ_0)	$4\pi \times 10^{-7}$ H m^{-1}
Stefan's constant (σ)	56.7×10^{-9} W m^{-2} K^{-4}
Boltzmann's constant (k)	13.8×10^{-24} J K^{-1}

Definition

The electron volt	$1 \text{ eV} \equiv 160 \times 10^{-21}$ J

Answers to Problems

1.1   1.4 μm.
1.2 $\sim 50 \times 10^9 \, m^{-3}$.
1.3   16.
1.4   Based on further reading by the student.
1.5   Based on further reading by the student.

2.1 387×10^{-21} J. 2.49 eV.
2.2   414 nm.
2.3 16×10^{-18} J or 100 eV.
2.4 ~ 3.
2.5   12.1 eV, 102 nm.
2.6 $3.9 \times 10^{21} \, s^{-1}$, $4.7 \, \mu J \, m^{-3}$, 400×10^{24} W.
2.7   5800 K, 500 nm.
2.8 7.8×10^{-18}.
2.9   600 MHz, 790 000.
2.10 633 nm, 1.3 pm, $0.03 \, cm^{-1}$.
2.11 $2 \, m^{-1}$.
2.12 0.78.

3.1   632.8 nm, 668.0 nm.
3.2 $6.2 \times 10^{18} \, s^{-1}$.
3.3   497 nm, 691 nm.
3.4   10 MW, 800×10^{15}.
3.5   38.2 mJ.
3.6 ~ 260 J.
3.7   Procedure as per Worked Example 3.4.
3.8 $15 \times 10^{15} \, m^{-3}$.
3.9   Procedure as per Worked Example 3.5.

4.1   42 mlm.
4.2   Prefer a detector with peak response near 1 μm. For measurement of total energy, a thermal detector is most suitable. For display of spectral profile, use a detector with ~ 1 ns risetime needed, say pin or avalanche.
4.3   For simultaneous measurement of flux and pulse duration, you need a pin or avalanche diode. Incident energy should be reduced using filters to avoid damage to the detector.
4.4   25 μA.
4.5   3.4 V.

5.1   3.9 mW, 156×10^{-6}.
5.2   100 μW.
5.3   0.95 μW.
5.4 $- 22.7$ dB, 426 nW.

5.5 ~9 nA.
5.6 2.8 mW sr^{-1}.
5.7 57 mm, 0.14.
5.8 160 nW.

6.1 Based on reading of text.
6.2 Based on reading of text.
6.3 1.523.
6.4 Based on reading of text.
6.5 ~300 photons (depending on assumptions made).
6.6 9 m.
6.7 Based on reading of text.
6.8 1.6 km.

Index